Transcending Space

Transcending Space

Architectural Places in Works by Henry David Thoreau, E. E. Cummings, and John Barth

Taimi Olsen

Lewisburg
Bucknell University Press
London: Associated University Presses

Associated University Presses
440 Forsgate Drive
Cranbury, NJ 08512

Associated University Presses
16 Barter Street
London WC1A 2AH, England

Associated University Presses
P.O. Box 338, Port Credit
Mississauga, Ontario
Canada L5G 4L8

The paper used in this publication meets the requirements of the American National Standard for Permanence of Paper for Printed Library Materials Z39.48-1984.

Library of Congress Cataloging-in-Publication Data

Olsen, Taimi Anne.
Transcending space : architectural places in works by Henry David Thoreau, E. E. Cummings, and John Barth / Taimi Olsen.
p. cm.
Includes bibliographical references (p.) and index.
ISBN 0-8387-5401-5 (alk. paper)
1. American literature—History and criticism. 2. Architecture in literature. 3. Cummings, E. E. (Edward Estlin), 1894–1962—Knowledge—Architecture. 4. Thoreau, Henry David, 1817–1862—Knowledge—Architecture. 5. Cummings, E. E. (Edward Estlin), 1894–1962—Settings. 6. Thoreau, Henry David, 1817–1862—Settings. 7. Barth, John—Knowledge—Architecture. 8. Space and time in literature. 9. Architecture and literature. 10. Barth, John—Settings. 11. Setting (Literature) I. Title.
PS228.A74 O47 2000
810.9'357—dc21 99-054649

PRINTED IN THE UNITED STATES OF AMERICA

Contents

Illustrations

Preface

If we take place *in the narrow sense of a physical envelope for the self, and consider the books which try to provide as literal a shelter as possible, we trace the specially spatial imagination expressing itself in words as it does in buildings . . .*

—Robert Harbison, *Eccentric Spaces*

THIS STUDY CONCERNS THE FUSION OF TRANSCENDENT IDEAS OF SPATIALITY onto one of the most rigid demarcations of normal space—the room. The dynamic transformation of static, solid structures, made up of flat planes, straight lines, and right angles into transcendent space which allows for expansive perception and self-perception tests the limits of the American spatial imagination. Three central works, *Walden*, *The Enormous Room*, and *Lost in the Funhouse*, serve as case studies of the metonymic use of architectural structures in literature and demonstrate the symbiotic connection between literary space and physical, or "real," space.

Examples of the fascination with architecture in American literature come to mind in the works such of nineteenth-century authors as Henry James, Edith Wharton, and Edgar Allan Poe; these texts often entail a confrontation with Poe-esque confinement to a house, as in James's *The Turn of the Screw*. In this study I explore a divergent trend in which authors examine their structural surroundings and transform these containers, search for gaps, breadth, and openness as they begin to define physical and spiritual freedom. By challenging preconceptions of structures as solid and permanent, Thoreau, Cummings, and Barth find a sense of the infinite in unlikely places, enlarging our understanding of the physical and cultural function of space. In 1947 Charles Olson declared "I take SPACE to be the central fact to man born in America,"[1] and the particular spatial imagination of Americans influences how we might construct our external and internal environment.

"Space still tends to be treated as fixed, dead, undialectical; time as richness, life, dialectic," complains Edward Soja in *Postmodern Geographies*.[2] One goal of this study is to refocus our understanding of American literature to include the exploration of spatiality in American texts. The critical apparatus of this study centers on two texts: Michel Foucault's essay "Des Espaces Autres" ("Of Other Spaces") and Henri Lefebvre's *The Production of Space*. Architectural criticism and histories, particularly Bruno Zevi's *Architecture as Space* and articles by behaviorist architects such as Jon Lang, also provide an important groundwork in tracing the American spatial imagination over two centuries.

The second goal is to examine the relationship between the physical and literary construction of spaces. Michel Foucault's essay serves as a linchpin for this topic. He identifies special sites called "heterotopias" that function as physical places yet also serve distinct cultural functions: "There are also, probably in every culture . . . real places, —places that do exist and that are formed in the very founding of society—which are something like counter-sites . . . in which the real sites, all the other real sites that can be found within the culture, are simultaneously represented, contested, and inverted."[3] His examples of such special sites include hospitals, theaters, museums, and brothels. I add a new category: the transcendental heterotopia, a place that challenges our traditionally static and rigid understanding of space. This type of site fluctuates dynamically between externalized and internalized space.

I use the term transcendentalism less to describe the historical movement than to signify those authors influenced by that movement and concerned with transcendence as a spiritual state. In the last chapters the discussion returns to Kant and the concept of origins and a transcendent "I" in order to argue against the privileging of one facet of the transcendent experience, or one aspect of a spatial field, over the other. Thoreau, more than Emerson, struggled to nurture a transcendent vision *within* the physical world instead of attempting to see *through* and beyond the world as Emerson did. Joan Burbick notes that Thoreau "came up against an unnerving sense of the phenomenal world that often challenged many of the assumptions of transcendentalism."[4] In *Walden*, his solution is in part to meld transcendentalism to everyday existence, thereby transforming his home into a heterotopia. He explores the implications of such a site through language that alternates between real experience and idea, between soil and preconceptions. "Let us settle ourselves, and work and wedge our feet downward through the mud and slush of opinion, and prejudice, and tradition, and delusion, . . . till we come to a hard bottom and rocks in place, which we can call *reality*."[5] Thoreau had a surprisingly fine-tuned sense of space, one that anticipated the turn-of-the-century explorations of microscopic and cosmic spaces. His understanding of space includes the metonymic with the metaphoric; when he describes rocks and trees, he means real trees, as well as their symbolic counterparts. "This world has many rings, like Saturn, and we live now on the outmost of them all . . . What do botanists know? Our lives should go between the lichen and the bark" (312). By redefining the perimeters of space, he challenges us to effectively reinterpret our environment.

This study also examines the use of language structure as a counterpart to architectural structure. In considering the parallels of these two structural systems, Cummings and Barth provide strong examples of the use of experimental language to describe transformative spaces. When Cummings finds himself in a French Detainment Center during World War I, he changes the nature of imprisonment. As Lefebvre would suggest, Cummings *produces* space in his "enormous room," a place which he describes as "suddenly and entirely alive."[6] The production of space relies on the movement between and interaction of elements in space. I argue that the space on the page replicates in miniature the movement between elements in an environmental space.

Most importantly, this study will examine the pervasive influence in the spatial imagination of dynamism (a term used particularly during the Futurist movement). The fluidity of space and the interaction of forces in space constitute a relatively new and exciting field of inquiry for architects and critical theorists alike. In a review entitled "The New Architecture," Ada Louise Huxtable describes the reconception of space by contemporary architects as embracing profound changes in spatial concepts: "Architecture is no longer conceived as the making of a formal "container," as it has developed over centuries of stylistic evolution. These [new] architects think first of the interior space and second of the enclosure; they handle space not as a finite form, but as fluid and open-ended; for the user, it becomes a serial rather than a static event."[7] The observer becomes a creator too of "fluid" space, through the body's perceptions and movements as that person interacts with an architectural structure. The authors of transcendent spaces understood this aspect of space implicitly. They describe interactions which strip place of its instituted function as a container for the body and instead find accommodation for the spirit.

As a consequence of the organizational method of this study, the parallel histories of literature, architecture, and science connect and reconnect at various points. This approach reveals points of exchange between these fields but does not give the reader a complete linear history of any one field. In my consideration of Barth, these fields meet again through his interest in kinetics. His recent lecture on chaos theory gives rise to questions of his own representations of space. Although Barth's fascination with new scientific developments leads us away from normal, human-scaled space to a consideration of micro and macro systems, I agree with Barth that in literature, these concepts are most useful metaphorically. Recent spatial theories rely on complex calculations involving dimensions beyond the four of Einstein's space-time continuum. So far, scientists have not been able to normalize completely the mathematical computations involved. However, new theories encourage our sense of spatial dynamism and its continued influence on other fields of thought.

Many people have contributed time and support to the writing of this book, and I wish I could thank all of them here. I would like to thank William Harmon, particularly for his delight in unorthodox projects and his confidence *in media res*. Linda Wagner-Martin offered years of encouragement and suggestions as well; her mentoring is much appreciated. Thank you to Philip Gura and Townsend Ludington for suggestions on various parts of the manuscript. My thanks also to Norman Friedman, director of the E. E. Cummings Society and editor of *Spring: The Journal of the E. E. Cummings Society*, for his example as a Cummings scholar and his encouragement of my own research and writing about Cummings. Finally, I am grateful for supportive parents, Anne and Blayne Olsen, and most of all for the generous aid of my husband, Thomas Beeson.

I am grateful to the authors, editors, and publishers for permission to reprint the following: Lines from "irreproachable ladies firmly lewd" (c) 1923, 1925, 1951, 1953, (c) 1991 by the Trustees for the E. E. Cummings Trust. (c) 1976 by George James Firmage, from COMPLETE POEMS: 1904–1962 by E. E. Cummings. Edited by George J. Firmage. Reprinted by permission of Liveright Publishing Corp. Parts of chapters two and three of this book originally appear as: "Language and Silence in The Enormous Room," *Spring: the Journal of the E. E. Cummings Society*, New Series 1 (October, 1992) and "E. E. Cummings and the Futurist Art Movement," *Spring: the Journal of the E. E. Cummings Society*, New Series 5 (October, 1996). Reprinted by permission of the editor, Norman Friedman.

Transcending Space

1
Introduction: Architectural Space

IMAGINE THE INNER SIDES OF A BOX, A COFFIN, A ROOM, A HOUSE, ANY walled space, and you will likely associate those spaces with containment. Descriptions of such places, when emphasizing their restrictiveness, can often invoke our fears of entrapment and death. Recall Edgar Allan Poe's multiple invocations of burial alive. In "The Pit and the Pendulum," the narrator describes how, in his dark prison, "I thrust my arms wildly above and around me in all directions. I felt nothing; yet dreaded to move a step, lest I should be impeded by the walls of a tomb."[1] Ralph Ellison's invisible man speaks to us from an underground room, his hideout and headquarters, insanely lit with a thousand lightbulbs. The plot of *One Flew Over the Cuckoo's Nest* revolves around escape from imprisonment in a mental hospital secured with barred windows, locked doors, and fences. These images resonate within us and, in human history, find a powerful metaphor in the medieval conception of the body as the soul's prison. By the seventeenth century, in "The Dialogue Between Body and Soul," Andrew Marvell describes the body and soul as mutually imprisoning, using architectural images to evoke physical sensations of destructive entrapment:

SOUL

Oh, who shall from this dungeon raise
A soul, enslaved so many ways,
With bolts of bones, that fettered stands
In feet, and manacled in hands. . . .

BODY

What but a soul could have the wit
To build me up for sin so fit?
So architects do square and hew,
Green trees that in the forest grew.[2]

In these stanzas the tension between the soul and body exists in opposing metaphors: the dungeon is subterranean and solid, while its instruments recreate fears of restraint; the construction site, dead and deceptively open, brings associated feelings of frustration and hopelessness. We often expect to be presented with this tension when confronted with images of containment, images which continuously pull at our imagination. Postmodern literature follows this tradition with paranoid fictions of entrapment and conspiracy. Fredric Jameson proposes that "successful spatial representation" need not be "uplifting" but can be seen as a "narrative of defeat"[3] and, in so saying, recognizes the dominant trend in how we represent our reality.

I wish to invert this viewpoint and look at an antithetical trend in American literature that describes the transformation of enclosed spaces. Authors challenge us to revise our habits of thought by remaking images of containment, examining structures for gaps, breadth, and openness. To a large degree, this trend begins with the American Transcendentalists. In his *Essays*, Ralph Waldo Emerson repeatedly refers to the power of the soul to break free of both metaphoric and real walls. "[T]he walls of time and space have come to look real and insurmountable . . . Yet time and space are but inverse measures of the force of the soul."[4] For the Transcendentalists, to be physically and spiritually free from limitations and confinement, one can rely on the power of the soul.

While the Romantics sought open fields and decaying picturesque cottages as spurs to their imagination, Emerson aspired to open spaces as opportunities for the release from ego. Walking across a commons, he describes "my head bathed by the blithe air and uplifted into infinite space . . . I am the lover of uncontained and immortal beauty."[5] The body precedes the spirit into a union with the immortal. When in 1947 Charles Olson declared "I take SPACE to be the central fact to man born in America," he suggested that the geography of the States—with its open plains in the Midwest and expansive oceans off both coasts—determines American character. Yet while many American authors traditionally focus on the peculiar, open, and diverse topography of the states, other authors have reconstructed the dialogue on place and re-envision enclosed places. Henry David Thoreau's *Walden* is one example of "an obsession in American literature with plans and efforts to build houses, to appropriate space to one's desires."[6] Houses are a common prop in literature, but how many authors bring the house to the center of the text and reconstruct the meaning of interior space?

Richard Poirier assumes logically, as do most critics, that space can be controlled by our desires. We are just coming to realize that space is not appropriated by any one force but is constructed out of the conjuncture of many forces—builder, resident, neighbor, environment—and that each new element propels the ongoing metamorphosis of space. The Transcendentalists foresaw the dynamism of these interactive forces. Thoreau shows how building a house opens space, and he writes about the need to reconfigure social preconceptions of dwelling into an understanding of the dynamic quality of space. As with the other authors considered in this study—E. E. Cummings and John Barth—Thoreau locates a physical place for himself where he can consider the nature of four-dimensional space, the space we live in.

The authors included in this study focus on the expression of the infinite in unlikely places as they challenge preconceptions of structures as permanent and solid. They show us that the division between exterior and interior need not be abrupt and definitive, and that 'crossing the threshold' can be a liberating rather than a dangerous experience. Constructed spaces need not be considered static if we refocus our attention on our active perceptions of and involvement in our environment.

By using the concrete image of a room, an author can explore questions of structure, containment, stasis and dynamism, experience and perception—all of which can play upon deep-seated convictions about rooms. The room has a long literary history as a metaphor for conditions ranging from imprisonment to familial security. This history guarantees the image's resonant quality, which in turn allows the writer greater descriptive freedom. Gertrude Stein's meditation on rooms in *Tender Buttons* covers disparate topics—gender issues ("Startling a starving husband is not disagreeable"), soldiers and warfare (including references to wartime occupation, victory, and mercenaries), and religion, to name just three.[7] When writers combine descriptively powerful images of interior space with transformative experiences in their own lives, the spatial imagination opens up like reflecting mirrors in a funhouse, showing space to be fluid and changeable.

Both critical theory and architectural aesthetics have conjoined lately in an interest in the variables of human conceptions of our space. Both disciplines have explored the potential, individualized meanings (functional and symbolic) of particular places such as the home or office building. Out of their interest in the dynamics of human spaces, Michel Foucault and Edward Soja point to the bias in critical studies against space. Soja complains that "[s]pace still tends to be treated as fixed, dead, undialectical; time as richness, life, dialectic."[8] Architectural critics deplore the inadequacy of architectural plans and models to predict how the space will be perceived and used by the occupants.[9] Architect Jon Lang explains that he wants to focus more attention in classroom and boardroom design on four-dimensional reality rather than on two-dimensional planning: "Movement through the environment can be passive or active and can be at very different speeds. We attend to different types of information in the optic array [what we see] when we travel through it at different speeds or with different purposes. . . . The sequence of experiences becomes important, and the way the optic array transforms and the way certain elements remain the same are important."[10] A two-dimensional drawing or aerial photograph or even an architectural model does not accurately predict what a person perceives in the various places of a building. Architects not only want the *plan* to reflect fluid space but want the outcome, the *building*, to "take experience into account . . . a beginning of a new process."[11]

The idea of a spatial "environment" is also being examined and redefined by architects and urban planners who are replacing static concepts with dynamic models which represent the viewer as much as the maker. If we visualize the potential variety of perceptual experience in a given structure as a "web" or site of interrelated elements, then traditional valuations of basic design units (the dot, line, and plane) become less relevant. An interior space can be described not simply in terms of three dimensions but as a four-dimensional form, which architectural critic Bruno Zevi defines as the experiences of a person moving through space.[12] A two- or three-dimensional architectural model emphasizes form and privileges the structure of the "container"; an architect who also considers movement adds experiential aspects to the plan.

Our habit of thought leads us to conceive of interior spaces as static entities (a "room" has a predictable dimensional shape); Zevi and others encourage us to consider how personal experiences of interior spaces make those spaces dynamic. Given a long history of struggles to define space in terms of universal and elemental constitution, the idea that space and movement are inseparable, that space *is* whatever we have considered *in it*, that it is *both* "space" and "place," may be difficult. A conceptual history of space (very briefly) centers on self-conscious attempts at definition and is marked by periodic disagreements and heretical pronouncements. Copernicus and Galileo defined the idea of space as a matter of speculation, not faith; Descartes further specified space as an "extension" having the properties of height, length, and breadth. The popular notion of "Cartesian space," according to critical historian Cornelis Van de Ven, refers only in a limited way to two- and three-dimensional geometric regularity. Instead, Descartes' concept of space "should be interpreted . . . as an existential three-dimensional expansion of a corporeal reality." Modernist architects "interpret[ed] form as an expression of its inner content," and De Stijl artist Piet Mondrian conceived of "space itself as the visible immaterialization of form."[13] Postmodern architects often fuse architectural, geographical (topological), and technical notions of space with quantum physics. Writes Brian Massumi, "[o]ur existence is an ongoing topological transformation of a complexifying abstract ontological surface . . . The confound of light envelops form, and with form it envelops space, at which point everything becomes movement."[14] In the debate between form and function, function (not to be confused with functionalism), redefined as motion and use, surpasses form:

> Stable forms can be designed to interact dynamically, as bodies move past or through them singly or in crowds, or as sounds mute or reverberate, or as relations of surface and volume change with the time of day or season, or as materials change state with levels of moisture or temperature, or as the connection between inside and outside varies as an overall effect of these variations in concert with the rhythms of activity pulsing the city or countryside as a whole.[15]

Not only has the twentieth-century, popular notion of space thoroughly incorporated the "fourth dimension," but the traditional conceptual divide between "inside" and "outside" (space as contained by a container) is collapsing.

Gertrude Stein writes that "in every space there is a hint of more" as a way of describing the expansive quality of interior space that occurs when the observer interacts with the entity (space) (72). The walls may be a certain width apart, but a person's perceptions enlarge or reduce them. If the walls are curved in any way, then "width" becomes harder to assess. Plus, the difficulty of capturing the paradoxes of expanding space and movable structures is limited by our focus on the shape of the container. Instead of imagining space as an entity, we shut it out of our minds like the inner room of a Greek temple whose interior is completely enclosed and inaccessible, lacking any door or window.[16] We think we know about rooms.

Given our role as perceivers, individuals are more than passive receptors of information. The individual moves through and experiences space, then internalizes and personalizes perceptions, a process which architectural theorists and psychologists term "cognitive mapping" and discuss in terms of creating mental geographies. Cognitive mapping originated from Gestalt theories, which proposed that information comes to us first as a whole (or gestalt) in which we recognize forms and patterns before picking out details. It suggests that we carry around knowledge of whole environments in our head, but that we define "environment" in various ways, not with the strict accuracy of a cartographer. According to experiments in cognitive mapping, when people were asked to describe their home towns they gave priority to those houses, stores, and roads which they used most and had particular meaning to them.[17] Cultural constructs of "place" seek to impose names and characteristics on areas that are open to infinitely multiple perceptions. Space is fluid; there are always "hints" of other views.

While architects hope to develop new designs based on knowledge of how cognitive mapping works, such critical theorists as Foucault, Soja, Lefebvre, and others have looked at aesthetic and cultural concepts of place in terms of perceptions and patterning. (The topic of space attracts theorists from backgrounds as diverse as Marxism, Structuralism, Poststructuralism, Phenomenology, and Semiotics.) Architectural and critical theory overlap in looking at the ways people create meaningful spaces. Pearson and Richards in their introduction to *Architecture and Order* describe how individuals infuse meaning into places by taking active roles in naming, categorizing, delimiting, and valuing places: "[O]ur environment exists in terms of our actions and meanings; it is an existential space which is neither external object nor internal experience. . . . Space is perceived only as places. The environment is categorized and named. Through the cultural artifact of a name, undifferentiated space is transformed into marked and delimited *place*."[18] In a more radical formulation of this idea, one which posits creative power in the person experiencing place, Henri Lefebvre describes space as produced by the body: "each living body *is* space and *has* its space: it produces itself in space and it also produces that space."[19] By walking into a room we center ourselves (taking ourselves as center) and create a perspective of that room which determines its function; this perspective changes according to our lived experience in that room. For example, to use a popular 1950s stereotype, a housewife with children may see her home as a place of work, and perhaps as a symbol of limitations on her time; her husband might see home as a refuge *from* work, a place of leisure. Either party could change his or her daily activities and perceptions and thus change the nature of their shared space.

As his title *The Production of Space* suggests, Lefebvre emphasizes production as opposed to finished product because space is never a void; space always involves content and "it never quite emancipates itself from activity, from use, from need, from 'social being.' "[20] The "center" of creation is involved in a dynamic relationship with space, so that the idea of a fixed "center" is somewhat misleading. A living body necessarily participates in continual activities, relationships, and usages of raw materials; *how* we participate is a question that requires a heightened awareness of these metaprocesses. Charles Olson says of Melville that "what counts, Melville had, the *experience*, what lies under. And his own *force* to resolve the forces,"[21] which is one way to express the mythic American power of self-determination and progress. It also expresses (in the extreme) the idea of each perspective center as a force among forces. The idea of space as perceptual, as created and determined by the involvement of the perceiver, means that we do not need to look for an 'escape' from a confined space; we can transform space itself.

Emerson saw possibilities for heightened awareness in humanity's "insatiable desire" to "be surprised out of our propriety" (178). Authors who try to "surprise" us fight against cultural conceptions of our bodies and our buildings as confinements and repositories rather than places for transcendent experiences. Habits of thought,

Zevi explains, influence our perceptions of our surroundings. In order to develop a "new feeling for structure," we must accept new formulations and techniques before these techniques are used in construction. To illustrate the difficulty of mental shifts, Zevi describes the workmen's reaction to Frank Lloyd Wright's house "Fallingwater":

> The story goes that when the moment arrived to pull down the last support of the scaffolding form in which the great cantilevered terrace had been cast, the workmen refused to carry out the job. . . . [T]hey were not prepared to pay insurance to the families of the two men who would certainly be buried under the debris of that architectural "lunacy." When Wright, enraged, seized an axe and attacked the support by himself, some of the workmen fled. Yet the terrace . . . is still intact. . . .[22]

The "habits of thought" imbedded in the culture give rise to preconceptions not only of architectural structure but other structural contexts as well. The introduction of new structural concepts in any field provokes strong reactions from the public. Cubism dramatically broke through the conception of art as portraying form through three-dimensionality and replaced that conception with a new understanding of four-dimensional representation. In his 1913 study of Cubism, Guillaume Apollinaire remarks that the painters picked up on current scientific trends and were "led quite naturally . . . to preoccupy themselves with new possibilities of spatial measurement . . . designated by the term: the fourth dimension."[23] Cubists can represent a face from front, back, and both sides simultaneously, showing us different perspectives of that face which in reality the viewer would need to move to see. Poet-painter E. E. Cummings conceives of this technique as "seeing around" an object.[24] In his writings he reconfigures Cubist ideas through playing with the structure of language. In the following excerpt Cummings manipulates syntax and word choice to represent angles of vision:

> irreproachable ladies firmly lewd
> on dangerous slabs of tilting din whose
> mouths distinctly walk
>   your smiles accuse

the dusk with an untimid svelte subdued
magic
    while in your eyes there lives
a green egyptian noise[25]

A cubist painter might represent this picture of women walking down a sidewalk by twisting their limbs, enlarging their smiles, and coloring their eyes vivid green. Additional sensory modes pervade the space of the poem, suggested in descriptions of noise, vibration, and propelled bodies.

In other poems Cummings captures the experience of interior spaces, as when he describes McSorley's bar through noises, atmosphere, surface textures, movements, and shapes: "Inside snug and evil. the slobbering walls filthily push witless creases of screaming warmth chuck pillows are noise funnily swallows" (110). The anthropomorphized room conveys a drunkard's perspective; loosely structured sentences and unusual syntax emphasizing recombinations of words (as in "screaming warmth") recreate the feeling of disorientation. Cummings grounds his poem in a real place announced in the first line ("i was sitting in mcsorley's"), a bar he frequented when living in Greenwich Village. By describing qualities of that space through sensory experience, he transforms the solidity of the room. The poem expresses real aspects of the bar scene, made possible through the interaction of authorial perspective and real material.

Michel Foucault's concept of heterotopias, sites with special functions, is of particular relevance to this study. Foucault defines a particular type of place that melds physicality of place with cultural constructs (the expectations one has for that place). In his lecture "Des Espaces Autres" ("Of Other Spaces"), he explains that heterotopias are real sites, physical places which challenge the meaning of other sites:

> [They] are something like counter-sites, a kind of effectively enacted utopia in which the real sites, all the other real sites that can be found within the culture, are simultaneously represented, contested, and inverted. Places of this kind are outside of all places, even though it may be possible to indicate their location in reality. . . . these places are absolutely different from all the sites that they reflect and speak about. . . .[26]

Because of their special cultural functions, heterotopias occupy unique places in cultural history. Foucault identifies the roots of heterotopias in places of crisis (hospitals, retirement homes) and places of deviance (museums, fairgrounds, gardens, cinemas, and brothels). The idea of a museum brings to mind particular qualities and expectations about its purpose and function that differ completely from the generic image of the house. Museums deviate from ordinary places by being set aside and marked for a special function. Crisis heterotopias, also set aside and flagged in some way, serve people in some "crisis" such as mental or physical illness.[27]

The interaction of external and internal realities in a heterotopia occurs during experiential perception, practical usage, and imaginative interpretation of a given structural place. Crisis and deviance play a role in the origination of all heterotopias. Foucault concerns himself with places which have a special function in society and which are recognized by a community (much as a church is set apart and designated for spiritual matters). In order for a structure to be "marked" by society as having a peculiar function, that place must be separated from and deviate from the established norms of society. McSorley's bar belongs to a heterotopic category since it is a city bar, a place set aside for men (primarily) to drink. This study adds a third category, heterotopias of transcendence, which encompasses those places that have been recreated by an individual and then shared in some way with others (as through writing). These transformed places have already traversed in some way the crisis of separation from societal norms. Such a place embodies conundrums, since it is both fixed and dynamic; it serves as a sacred place for that individual, a place of power. The person in a transcendent place has survived a spiritual crisis and his or her "deviance" is actually the fulfillment of a widespread cultural desire to attain 'enlightenment' and freedom. A heterotopia of transcendence is a real space which provides "compensation" for all other mundane spaces; it is also "a space of illusion that exposes every real space . . . as still more illusory" and so fits Foucault's definition.[28] One purpose of this study is to discuss places such as Walden, the detention camp at La Ferté-Macé and the Ocean City Funhouse as heterotopias.

Foucault defines "heterotopia" in terms of spatial interconnections within a site, although he does not consider the kinetic nature of those connections. He comes close to revealing the dynamic quality of "sites" when discussing the (presumably active) connections between the elements of a site. "The site is defined by relations of proximity between points or elements; formally, we can describe these relations as series, trees, or grids. . . . Our epoch is one in which space takes for us the form of relations among sites."[29] Foucault envisions a site as a web of pathways. Just as an artist can focus on positive or negative space, we can examine the structure of the path or the possibilities presented by that path. For instance, Henry James in "The Jolly Corner" shifts his attention from the containment structure of the house (one of many "builded voids" in New York City) to the pathways in the house. Brydon moves around the empty house "rejoicing above all, as much as he might, in open vistas, reaches of communication between rooms and by passages."[30] James describes Brydon's perceptions and his imagination of the past as symptomatic of an extreme mental state, yet the idea that a building's purpose lies in communication captures the nature of space as an electric medium.

The power of transcendent space lies in the use of space as an interconnected site. Even Foucault has difficulty registering the mobility of space and the play of forces connecting on a physical plane. I will mention in passing Gertrude Stein's consideration of structural issues, since her representation of movement in the "Rooms" section of *Tender Buttons* shows her prescience of later architectural views of space as a dynamic field. "Rooms" presents us with an interior space which Stein describes through commonplace images of furniture, food, and clothing. Her descriptions rely on standard devices of description of color, size, number, light, and shading. Stein's style forgoes other conventions of standard description: linear coherence in describing the room's contents, linear movement around the room(s) as it is described, the grouping of objects logically (describing the contents of one room before moving to another room), and restriction of description to singular instances (mentioning one object once unless it is significant to the story). Instead, Stein's description relies on complex connections and associations to provide the reader with a description of dynamic space: "Act so that there is no use in a centre. A wide action is not a width. A preparation is given to the ones preparing. They do not eat who mention silver and sweet. There was an occupation" (63). Stein concludes that "a wideness makes an active center" (77). Contrary to traditional views of humanity's need to maintain a stable center, Stein posits a space that is not centered but rather is filled with action and occupation immeasurable in three dimensional terms ("action is not a width"). Her choices in language emphasize the nature of the room as a dynamic interior space interacting with its occupants. This passage focuses on the triad of act / action / active. The use of three syntactic forms—noun, verb, and adjective—represents the various ways which we experience life as movement, object, and quality. In this way Stein defines space as experience.

Stein's refiguring of the space of the page compliments her reconsideration of traditional constructs about everyday space. The space of a text can be described as a site, or series of sites (as I will discuss shortly, both language theorists and architects employ the term "site"—not necessarily in the same manner but with consistent similarities). Language can be conceived in terms like that of Foucault's grid; whether spoken dialogue or written text, it forms in our mind out of a series of connected elements.[31] These related elements are built out of one's mental lexicon, using one's sense of the syntactical patterning in language. The mind, in forming a sentence, moves through paradigmatic (associative) and syntagmatic (syntactical) choices. In the passage cited above, Stein uses rhyme and alliteration to form other connections within the paragraph (do / who and eat / sweet form rhyming pairs; the alliterative use of "s" and "w" also weaves patterns in that sentence). As Ferdinand de Saussure explains, syntagmatic, or horizontal, meaning governs the sentence through "a succession of meaning-units or words in time."[32] Paradigmatic meanings are located vertically, from outside the immediate context of the sentence in our mental dictionaries. As we move across the sentence (in speech or writing), we choose words through associative, mental lists. The combination of the horizontal and vertical creates a "space" of text in which the reader pulls meaning from multiple directions.[33] When reading written text, our eyes also move physically around the page (not simply left to right, as is assumed; a word in bold print, for example, might catch our eye first). Our mind also 'travels' through the mental space of meaning, as when words (images, metaphors, symbols) infer their etymological, mythic, and cultural history, and as these meanings react to each other spontaneously in the context of the sentence. Although Lefebvre devalues language for its occupation of a lesser, mental space, he ignores the primary impact of the physicality of written text and the associations upon which language can draw.

In formulating the simplistic question "are language and architecture alike?" (and thus determining if the reader can experience a real place secondhand through language), the trend to see everything in terms of "metaphor" has reduced the options of the critical thinker. Gérard Genette reverses this simplifying trend and reintroduces the concept of literary figures. In discussing the connections between architectural and language structures, the replacement of "metaphor" with the concept of a figure is useful. Language can describe containers such as rooms or houses, but it is not completely "like" the container it describes. In one sense language contains meaning, but the inner space of a word is mental rather than physical. It can not be architectural since the definition of architectural form depends on a structure's viable inner space.[34]

A heterotopia differs from a literary metaphor in that it combines the "real" physical world and cultural constructs, but it parallels special language sites that draw attention to and use the physical characteristics of language. One key similarity exists between heterotopias and language; as just mentioned, words can also function as "sites" since they bring meanings buried in etymologies to the site and create interconnections through syntactic structures and patterns of meaning, sound, and sight. Critics like Genette have used the image of language as a palimpsest to convey an understanding of the peculiar dimensionality of language and the depth of associations which a word or phrase can convey to the reader through the ambiguities of meaning. "Every figure is translatable, and bears its translation, transparently visible, like a watermark, or a palimpsest, beneath its apparent text."[35] The effect of light, watery layers of meaning underneath the printed text can be created through associations of sound and sight as well as meaning.

By using the forms and sounds of language and the meanings associated with words, a writer can create layers with depth and width, allowing for movement in multiple directions. The space that these associations occupy is highly variable. We do not literally find layers 'underneath' the words; meanings are located in the dimensional space of our brains. Sight associations, however, do exist on a plane since our eyes can move freely around the page. As Thoreau suggests, we are able to apprehend meaning "by the perpetual instilling and drenching of the reality that surrounds us" (141). For

Bachelard, a figurative image (like Genette's figure) "has an entity and a dynamism of its own" and analysis of that image requires moving "towards the outpourings of the mind and towards the profundities of the soul."[36] A word can encourage us to recall associations beyond the limitations of space, because of the author's understanding of the ways in which a word can resonate in the reader's mind. Poetic images come before thought and invoke a spiritual "inner light" which transforms the image into a dynamic entity;[37] they reverberate in our memory and imagination as an oneiric experience. Thus, the dynamic quality of language can be used to "transcend" perceived structural limitations.

Bachelard concentrates on images that inspire; this study will look at transcendent sites and suggest that the "prison-house of language" can be reconfigured through a process similar to oneirism, whereby words expand and encompass meaning. These "sites" involve combinations of image, figurative language, and structural components. The language site is then a figure of figures, a complex with interacting parts but without an overriding, central figure. A language site can involve metaphor, referential meaning, oneirism, rhyme and rhythm, syntax and punctuation—in short, all the mental and physical qualities of language. Conceiving and experiencing both literary and architectural structures pose similar challenges. How do we recognize and experience the dimensions of language? How does a writer (or an architect) describe unusual perspectives? How does the spatio-temporal dimensionality of language affect meaning (or of architecture affect design and outcome)? How does an author (architect) convey and encourage transcendent experiences?

It is a remarkable feat to create a three-dimensional house and then move beyond the prescribed three-dimensionality of that container to imply other experiences available to the character-occupants. In these dynamic sites, we are taken one step further beyond what Gaston Bachelard calls the "oneiric house" of the human self, therefore fulfilling his expectation that "by approaching the house images with care not to break up the solidarity of memory and imagination, we . . . touch the ultimate poetic depth of the space of the house. . . ."[38] The observer expands the meaning of both place and text through participation in that site. In our role as perceivers, then, we become co-creators.

The intended function of a site (a site burdened already by prescriptive assumptions about it) can pose obstacles that interfere with creative revisioning. As an example of the difficulty of creating a multidimensional, transcendent space, consider the uneasiness that Hawthorne expresses in the "Old Manse" where he wrote *Mosses from an Old Manse*. In his introduction to this short story collection, Hawthorne takes the reader on a "tour" of the house he has (in real life) rented, describing the house in terms of its functions for his writing self. He divides the manse and its grounds into spaces that serve different functions. The first floor, a semi-private, comfortable and welcoming space (he tells us that his guests regularly snooze on the sofa) contains a "nook" for study and writing. This floor represents both the public self and the working self. The river outside serves as a place where Hawthorne's imagination is "free" and where he navigates the physical and mental landscape in his canoe, surrounded by the stimulation of light, movement, and relatively open space. Next, Hawthorne invites us up to the garret, where we are "in the rationalized zone of intellectual projects."[39] This garret serves as a storehouse for the mind, where Hawthorne finds mostly dusty, outdated, and uninteresting religious works.

Hawthorne expresses disappointment and reservations about this house due to its history of sheltering clergymen; he worries that the house will not inspire his imagination. He finds comfort, however, in places such as his nook and the grounds. The garret, too, has a bright spot. In a corner, Hawthorne sees an attic living space for students, and their inscriptions on the whitewashed walls capture his imagination. The light from small windows refracts off of these "brief records and ejaculations"; of all the shadowy corners of the house, Hawthorne denotes this particular corner as a place of inspiration.[40] The daylight metaphorically suggests that the inscribed words are value-laden with a suggestive inner light. These inscriptions draw our imagination into a sensual experience of verticality, depth, and luminosity.

Hawthorne's hesitation about occupying a clerical house, his disappointment about the slim findings of writer's material in the garret, and his denigration of the worth of his "fitful sketches" (34) have been explained by critics in various ways, often as the author's pose designed to set the reader at ease. Yet something is uneasy in the Old Manse—it has a literary ghost in the form of Ralph Waldo Emerson. He intrudes in the introduction, and Hawthorne, like a dutiful tour guide, or a nervous writer, tells us that Emerson wrote *Nature* in the study (5). Later he describes Emerson, a "poet" and "original Thinker," wandering the countryside around the house (30–31). However, Hawthorne never invites Emerson into the oneiric house, perhaps because his "lofty thought" would overshadow his own "shallow" writing (31–32). Despite Hawthorne's justifications, Emerson remains outside, a spiritual element circling the house.

Emerson represents a dimensionality that Hawthorne could not or would not integrate into his house. Likewise, nineteenth-century transcendental thought presented new concepts of mental and spiritual space; Emerson's description of the oversoul or his expanding "circle" of experience suggests spatial and temporal dimensions. Emerson created an entire spatial map for his philosophy, one that includes a field of depth (through a three-dimensional concept of the oversoul) and of expansion (the circle). He used his spatial imagination to free his thought, unlike Hawthorne, who remained tied to the traditional duality of inside (containment) and outside (freedom). In his tour of the garret, Hawthorne almost expands his concept of space but fails finally to reconfigure his literary space.

Writers who invite Emerson into their houses grapple with the "phenomenology of the soul" which Bachelard finds in poetic images that 'reverberate' in the imagination.[41] These images move through the reader and convert what Emerson terms the "NOT ME"—body, house, and environment—into an interior part of our spiritual selves. Emerson explains that "Nature is not fixed but fluid. Spirit alters, molds, makes it. . . . [T]o pure spirit it is fluid, it is volatile, it is obedient. Every spirit builds itself a house, and beyond its house a world, and beyond its world a heaven" (56). Through our houses, we can "represent or symbolize what is tantalizingly unrepresentable.[42] Realizing the kinetic nature of solid structures changes the terms of the dialogue for Emerson and for us. The "ME" moves outward to explore the physical world.

Cummings' response to the implications of dynamism within transcendental thought was to play with language in order to evoke its expansive nature. He realized that language itself was fluid. The term "La Liberté" in *The Enormous Room* becomes a palimpsest as each reference to liberty provides an added layer of meaning, often ironic and sarcastic. The term is also a language "site" created by

Cummings' use of formal structure. A detainee nicknamed "The Bear" takes the word and tattoos it on his chest. Cummings draws our attention through enlarged, bold-faced, and capitalized print to the tattoo "VIVE LA LIBERTÉ." He also points to the irony of how this man has taken an idea and "scooped it out of its environment."[43] The "environment" consists mainly of the military establishment that prevents this man from mailing any letters; it might also refer to the greater cultural environment and France's revolutionary phrase "liberty, equality, fraternity."

Like Cummings' play with language dimensions, Gertrude Stein's conceptualization of space remains inseparable from the way she has written about her ideas. The interconnections of the "site" of her text with a "room" she writes about include plays on words and syntactical structures. Stein draws our attention to the stasis-motion dialectic of noun-verb pairs (act / action, preparation / preparing), while active and passive voice orient the subject of the text ("There was an occupation"), and allusions to the ritual use of space, such as food preparation, are set in opposition to unexpected and radical decrees about spatial relations ("Act so that there is no use in a centre").

Stein suggests that a person can widen a seemingly fixed space through dynamic movement. Likewise, any description of space (such descriptions necessarily isolate and select items to describe) is tantamount to establishing a center: "If the centre has the place then there is distribution. That is natural. There is a contradiction and naturally returning there comes to be both sides and the centre. That can be seen from the description" (64).

When we choose a central point of perspective, the center and sides "naturally" line up—but only after contradictory space redistributes itself to fit the momentary perspective. In the sentence, the phrase "both sides and the centre" balances the words across the fulcrum of a conjunction.

As the "Rooms" section demonstrates (and as does most of Stein's prose), the reader "moves" on a variety of planes, both mental and physical, when reading a text. These planes, to reiterate, occupy vertical and horizontal space (the syntactic and associative 'space' of language) and the entire space of the written page—over which our eyes and minds can travel in multiple directions. Writers can make the act of communication more substantial, so that the reader notices the writing itself, by expanding the functions of language and giving the written word some "breadth" and "depth." Because imagination plays a crucial role in the act of reading, we lose sight of the physicality of writing—its height, width, edges and curves. Written language occupies a real, two-dimensional space (really three-dimensional, considering that ink has volume), and although writing appears to move in a linear direction, words also move up and down (as with capitalization) and in circular motions (as with repetition). The reader literally "jumps around" the paragraph. In a sense, hypertext—in which clicking on a "linked" word literally changes the space of the page to introduce new text explaining associative and metaphorical meanings—is a literal (and comparatively limited) rendering of what language can already accomplish.

The physicality of Stein's prose is understated, yet it signals the reader to make mental connections to associated meanings, as in our understanding of ritualized daily activities. "Rooms" contains the world through references to domestic, religious, and governmental spheres; for Stein, an active person is occupied by space and instead of being contained, a person can choose their worlds. Noticing accidental combinations within space can give us a sense of reality: "A package and a filter and even a funnel, all this together makes a scene and supposing the question arises is hair curly . . . the whole special suddenness commences then, there is no delusion" (66). The reality comes not from a concrete object but from a *sudden event* in the midst of a spatial field. This point is also made on the level of the sentence, where the question is imbedded in a statement rather than separated with a question mark. As we consider the "scene" as "special," tracing back to further develop our mental image after the words have flagged the statement itself as "special," we wrestle with multiple interpretations of the three items listed. Do we picture a tableau of package, filter, and funnel on a kitchen table? Do we visualize items as representing progressive functions, from transmission to transcience? Do we visualize the acting out of these functions, the last being the most fluid? For the reader, Stein's rendering of the spatial field of a room offers experiences with and about structural dynamics.

In conceptualizing structural space, critics who consider structural systems of any kind to be inherently similar often draw metaphoric parallels between architecture and literature. One literary critic asserts that "any architect will verify that a house is a text."[44]

Yet, is there a difference between writing about a room and walking into one? Both media (written words and building materials) are concrete, located in space, handled systematically, and both are experienced through human perceptions and imagination. In the current mode of late twentieth-century thought, both systems are being redefined in terms of fluidity and dynamism. However, each structural system tries to meet the requirements established by its purpose (newspapers can be folded into hats but serve their intended function as reading material much more efficiently).

Any critical perspective that reduces all structures to one structural formation misses the complexity and interconnectedness of those structural systems, whether we are considering language, natural formations, or social constructs. Differences exist between walking into a room and reading about it; neither structural system can adequately convey the experiences of the other. When the focus is on process itself, we learn about how we experience structures and remain open to possible alternatives rather than being committed to one overarching structure.

Hypersurface architects would obviously suggest that in bringing electronic text and images together with architectural surfaces, they can combine the two systems of language and architecture and successfully explore the interconnections. We are in essence approaching the issue similarly but from different perspectives (whether incorporating text into architectural ideas or examining text through architectural thought). In hypersurface architecture, digitized media is fused with a dynamic, architectural surface. As Stephen Perrella explains, "[t]he term hypersurface is not a concept that contains meaning, but is an event; one with a material dimension."[45] Architectural projects find different ways to use architectural surfaces as 'media surfaces' with the rationale that there is already a need for such structures. "Information culture is spilling out into the built environment, creating a need for surfaces through which data may traverse."[46] These hypersurfaces are complex rather than reductive in that they serve multiple, dynamic functions. However, the critical language to evaluate the functionality of hypersurface architecture as both architecture and as communication remains essentially separate despite attempts to fuse the two systems. Perrella's use of the term "reader-bodies" places two functions side-by-side, recognizing the separate functions combined into one person (a person who both 'reads' and whose body interacts with surrounding structures).[47] Regardless, the interface between human body and lived experience invades most discussions of dynamic space.

Martin Heidegger is a philosopher for theorists of space, because he defines space through the life of the body. "[E]ven modern physics was compelled by the facts themselves to represent the spatial medium of cosmic space as a field-unity determined by the body as dynamic center."[48] As a result, we cannot discuss people and space separately; space "is neither an external object nor an inner experience."[49] Yet Heidegger defines dynamic space in a peculiarly passive manner. For him, a site gathers the landscape, and our buildings allow for that space. "Space is in essence that for which room has been made, that which is let into its bounds."[50] In this sense space converges on a center, the human body. However, the solid core of a center belies the dynamism of the "field-unity." Our bodies necessarily serve as the center of our perceptions, but they do not define all of space. Space consists of movable and interacting 'centers.'

In his 1953 foreword to Max Jammer's *Concepts of Space* (a foreword included in the expanded third edition), Albert Einstein joins in the process of rejecting absolute space and asserts that space can no longer be considered "empty." Instead, "[t]hat which constitutes the spatial character of reality is then simply the four-dimensionality of the field." The "components" of the field "depend on four space-time parameters."[51] Jammer remarks that this last definition of space by Einstein has never really come under criticism despite the ongoing philosophical debate between relativists and absolutists and the addition of new multidimensional theories of space (216).[52] Currently, theories of space deal with data from nonspatial and nontemporal microparticles and data on the cosmic scale; these theories typically work in dimensions higher than the fourth. This study does not claim to include these newest theories of space, choosing instead to remain focused on human-scale systems. The ideas of quantum physics about space, however, are influencing both literature and architecture, in postmodern literature's representations of "pluralistic" space or in topological architecture, which "imbricates temporal events-within form" in a space which is "an interconnected, dense web of particularities and singularities."[53] In both cases, physics has crossed over into other fields, leaving scientific dialogue and entering cultural construction. This study will end at this point, anticipating further discussion when proposals for the integration of digital technologies into housing structures and the computerized psyche posited by science fiction writer William Gibson conjoin in real, everyday space. So, given the current state of both fiction and architecture, the question remains, how do we transcend ordinary spatial constraints?

2

Transcending Space: *Walden*

If with closed ears and eyes I consult consciousness for a moment, immediately are all walls and barriers dissipated, earth rolls from under me, and I float. . . .

—Thoreau, *The Journal of Hery David Thoreau*

I

THOREAU, WRITING IN HIS 1838 JOURNAL, USES THE POWER OF HIS "CONsciousness" to break down physical barriers. For American transcendentalists the concept of consciousness indicated a heightened awareness of self stemming from Kant's views on the mind's centrality; Octavius Frothingham, a second-generation transcendentalist, notes that for Ralph Waldo Emerson, "the universe is the result of mind."[1] Kantian thought limits this power to shape the world; Kant felt that mind "cannot go out of itself, cannot transgress its own limits. It has no faculty by which it can perceive things as they are. . . ."[2] The language used by Emerson and other American transcendentalists suggests otherwise. As suggested by their writings, the mind interacts in real ways with the world, and the dynamics of the connection between mind and world point to interactions between the Emersonian "me" and "not me."[3] Thoreau intuits a complex interdependence between self and world in which he literally feels at home in nature.

Thoreau's notions of space present a puzzle since he rarely discusses space as such, and when he does he often uses metaphoric language. He acknowledges the space of intellectual work in his journal, citing the actuality of mental space: "For only absorbing employment prevails, succeeds, takes up space, occupies territory, determines the future of individuals and states, drives Kansas out of your head, and actually and permanently occupies the only desirable and free Kansas against all border ruffians."[4] Despite the solid dimensions suggested in his metaphor of maps, landscapes, and borders, Thoreau does not stray here from Kantian philosophy, nor suggest that inner reality can bridge over to outer reality. Thoreau remarks earlier that "to give the within outwardness, that is not easy."[5] In order to understand Thoreauvian space and to see whether or not Thoreau's barriers are real or metaphoric, we must understand that he bypasses Kant's notions of space to adopt Emerson's language of a dynamic relationship among soul, Nature, and Oversoul.

Kant identifies both time and space as fundamental, universal concepts preceding any experiential information or thought. "Space is an intuition, met within us a priori, antecedent to any perception of objects, a pure, not an empirical intuition."[6] Kant's space is a fixed entity, its self-evident properties deriving from Euclidean geometry, "considered to be the only true geometry of real space" for centuries.[7] Since space exists prior to perceptual experience, any fluctuations in our perception of space would be attributed to our perceptions rather than the nature of space. Space is "transcendental," one of "the fundamental conceptions, the universal and necessary judgments, which transcend the sphere of experience tributary to knowledge."[8] When Kant's followers relaxed this terminology, a gray area developed in that "[t]he disciples of Kant and Fichte assumed that man was a creative, recreative force, a being who had only to be conscious of the capacities within him to shape circumstances according to the pattern shown him on the Mount."[9] Aside from the nod to Christian determinism, Frothingham tells us that a "conscious" person can direct mental energies onto the world in a "creative" manner.

Emersonian self-reliance includes this ability to propel oneself to a "higher" state, a trajectory in which one's soul is aimed at the Oversoul. A person's intellect "is not like the vision of the eye, but is union with the things known."[10] The self reaches outward to experience the powerful mobility of a moving and expanding space. Although the material world may seem static to the unengaged eye, the person who becomes "immersed" breaks through this misperception. "Nature shows all things formed and bound . . . [yet the] intellect pierces the form, overleaps the wall, detects intrinsic likeness between remote things. . . ."[11] Emerson attributes a power to the self which can be attained by transcending the limitations of the self and joining in the Oversoul.

Catherine Albanese writes that "the Emersonian cosmology . . . was a world set in motion—and motion of a different sort from the peaceful rhythms in Eastern and Western correspondence until then."[12] Emerson proposes, in nonscientific terms, that the universe behaves differently when we change the nature of our expectations, including scientific assumptions: "And so, always, the soul's scale is one; the scale of the senses and the understanding is another. Before the revelations of the soul, Time, Space, and Nature shrink away. . . . The soul knows only the soul; the web of events is the flowing robe in which she is clothed."[13] The senses register the material world on a physical scale; the soul's scale both changes our perceptions and shows us a physically altered world. Emerson assumes that the scale of space relies on another, more ephemeral scale, and his previous assumptions about space easily fall away before the central power of the soul.

Although nineteenth-century thought assumed a fixed quality to space, American Transcendentalists prefigure twentieth-century concepts of space as dynamic and interactive, situated in the space-time continuum. Einstein summarizes three main concepts: 1) Aristotle conceived of space as place, defined as occupied by material objects. There is no empty space. 2) Copernicus and Newton's absolute space envisions space as a container, a three-dimensional volume of empty space. 3) Einstein developed a model of space as a four-dimensional field of four parameters (the fourth parameter is time).[14] As historian Cornelis Van de Ven asserts, architectural space is a late-nineteenth-century idea[15] and implicitly involves time and movement.

When literary scholars dismiss space as a static entity, they misdirect our interpretations of Thoreauvian space. Joan Burbick, who critiques and admires Thoreau's "vision of redeemed time," remains ambivalent towards Thoreau's space, with its "cartographical grids" which she identifies as three-dimensional configurations of landscape formatted through descriptive language.[16] She cites Thoreau's explorations of Walden Pond as an example of delimited space. Thoreau surveys the pond and concludes that "[w]hen I had mapped the pond by the scale of ten rods to an inch, and put down the soundings . . . I laid a rule on the map lengthwise, and then breadthwise, and found, to my surprise, that the line of greatest length intersected the line of greatest breadth exactly at the point of greatest depth. . . ." (552–53). Through measurements and descriptions of the deepest spot in Walden Pond, Thoreau recreates geometrical space for the reader. Yet Burbick does not point out that his scientific analysis presents us with a type of space we do not usually experience, even through firsthand observation of ponds. Often Thoreau presents his reader with an unusual conception of space chosen out of several possibilities.

Thoreau's "mapped perception of the pond [is] at once a static and transcendent view" (69). Burbick's own language undercuts the notion of the transcendent power of space with static terminology. In *Walden*, Thoreau creates a "cultural space in nature" that, in Burbick's words, is "bounded," "enclosing," "charted," "confined," and "condensed" with "contracted vistas."[17] Although she concedes that objects exist in a state of spatial flux, Burbick maintains that Thoreau "attempts to use the map as a form of representation that stabilizes space." With space stabilized, time freely provides the dynamics of life. By shutting out thoughts of time, "Thoreau is truly free to imagine transcendence, a place without change, and hence, death."[18] Perhaps unintentionally, Burbick uses static nouns to describe a "place" of transcendence, instead of emphasizing the process of transcending through an active verb form.

Thoreau, however, often imagines transcendence as dynamic and sets objects in motion: "The very globe continually transcends and translates itself, and becomes winged in its orbit. Even ice begins with delicate crystal leaves, as if it had flowed into moulds which the fronds of water plants have impressed on the watery mirror" (567). These sentences in *Walden* fall in the middle of his much cited "deep cut" passage. A bank of sand, cut to make a passage for the railroad, melts under a spring sun, and Thoreau identifies the vegetative forms suggested in the flowing sand and clay. The passing time allows for a transformation of solid earth; yet the forms themselves emerge not out of time but out of spatial configurations. This movement in space—the winged earth—creates a transcendent experience. Thoreau's favorite journal image, ice crystals, represents motion captured into a solid form. Once heat energy is applied to ice, the crystals melt. When ice cutters harvest the frozen Walden Pond, the blocks of ice they cut and lift out suggest this paradoxical quality. Each workman "cuts and saws the solid pond, unroofs the house of fishes, and carts off their very element and air. . . . It looks like solidified azure" (556). The solidity suggested by the large cubicles of ice fools the viewer into a conception of false space; the space inhabited by living beings, either human or fish, has no definitive borders and no timeless quality. It flows.

Time necessarily creates change and cannot be disassociated from space without creating a death-like effect and vice versa; yet critical opposition to considering the interdependence of time and space still exists between "the pious descendents of time and the determined inhabitants of space."[19] Edward Soja, commenting on Foucault's prediction of an "epoch of space," reminds the reader that critics continue to call for a reformation of the imbalanced prioritization of time over space, that if corrected would allow critical readers "to see spatiality with the same acute depth of vision that comes with a focus on *dureé*."[20] Soja uses a spatial metaphor purposively in this statement; his phrase, a "depth of vision," echoes one of Thoreau's favorite images used in *Walden* when he writes metaphorically that he would "wedge" and "burrow" to the solid foundation of the earth (400). Through his metaphorical language, Thoreau indicates his sensitivity to his body's interaction with space and its continual movement through space.

Geographer Yi-Fu Tuan distinguishes between "space" and "place" and in so doing, he also divides space and time: "From the security and stability of place, we are aware of the openness, freedom, and threat of space, and vice-versa. Furthermore, if we think of space as that which allows movement, then place is pause; each pause in movement makes it possible for location to be transformed into place."[21] People can only create necessary human places out of timelessness. Critics note that a major difference between *A Week on the Concord and Merrimack Rivers* and *Walden* lies in Thoreau's treatment of the journey motif in the former and his motif of a "sedes, a seat" in the latter (387). Themes of time, travel, history, and chronology dominate *A Week*, while in *Walden* Thoreau considers place, shelter, settlement, and stability, and compares the settler to the traveler. Yi-Fu Tuan, according to his dialectic of free, open space and secure, stable place, would consider Thoreau's experience at Walden to be an expression of that dualism in the contrast between house and nature.

Thoreau resists the metaphoric implications of settlement as immobility and death, and similarly resists the impasse assumed in Kantian thought between self and object. Instead he envisions a dynamic, creative link between self and the material world. Burbick points to Thoreau's understanding of perception as an ability that allows people to interact with the world. When Thoreau describes the pond from the view of "myriad eyes," he "describes the imagined vision of unrestricted perception, transcending the limits of time and space." These unique viewpoints "are not locked in the solipsism of the individual mind, but are also seen as a function of the object."[22] He *imagines* the pond from multiple viewpoints, not allowing his own limitations (of one viewpoint) to restrict his perceptions. "To myriad eyes suitably placed, the whole surface of the pond would be seen to shimmer . . . if there were as many eyes as angles presented by the waves, the whole surface would appear as bright as the moon."[23] Viewer and object participate in the act of perception. For Burbick, this passage describes Thoreau's understanding of the perceptual process. It also describes Thoreau's sense of mutable space, since the reflections he sees leap around like fireflies, or like small flames which "slide" back and forth in space. At the pond and other sites, Thoreau locates dynamism in seemingly static places by denying them their stasis.

Thoreau prefigures the turn-of-the-century thought of Bogdanov and other scientists when he intuits a link between object and perceiver. Because the perceiver always perceives within the context of a particular place at a particular time, Thoreau posits that the spatial configuration of a particular place necessarily affects the view, just as the conscious viewer brings his or her inner, recreative power to the site. Fifty years later, A. Bogdanov writes that space is "a form of social coordination of the experiences of different people."[24] Bogdanov posits that space adapts to human experience to create a "plurality of spaces."[25] Thoreau's consciousness of this constant interaction of space and body drives him to extremes of feeling: "Think of our life in nature,—daily to be shown matter, to come in contact with it,—rocks, trees, wind on our cheeks! the *solid* earth! the *actual* world! the *common sense*! *Contact! Contact! Who* are we? *Where* are we?" (646). In this passage from *The Maine Woods*, the primitiveness of nature overwhelms Thoreau's sense of self, yet the contact is necessary for mystical experience, linking spiritual awe to natural sites. Thoreau intuits a relationship between bodies and space that resides closer to an understanding of space as a dynamic field than as a unchangeable entity. Thoreau's sense of space includes an appreciation of movement through space and challenges our presumptions of spatial configuration even while he accounts for space through exacting measurement. His ascent of Mt. Ktaadn challenges the body's normal position in space:

> At first scrambling on all fours over the tops of ancient black spruce-trees . . . their tops flat and spreading, and their foliage blue. . . . I walked some good rods erect upon the tops of these trees, which were overgrown with moss and mountain-cranberries . . . Once, slumping through, I looked down ten feet, into a dark and cavernous region, and saw the stem of a spruce, on whose top I stood. . . . This was the sort of garden I made my way over . . . certainly the most treacherous and porous country I have ever travelled. (639)

Thoreau walks a doubled height, the height of the mountain peak and of the treetops. He sounds the depth of his path as he would measure the depth of the pond. In this paragraph he emphasizes directional movement in space: up, across, over, down. His porous path calls into question the nature of foundation, a word used in *Walden* to signify spiritual grounding. Instead of walking through an edenic garden, he walks over it. His position and movement are equally radical; once on the mountain top, he sees a familiar sight, Maine country "waving, flowing, rippling" (638). The landscape undulates, and his hike seems especially treacherous, because he has turned away from stability and taken on the challenge of dynamic space.

II

In his writings Thoreau frequently contradicts preconceptions of space. On a walk with William Ellery Channing, Thoreau sees a man in a carriage, crossing a river by an arched bridge, and notices that "the stratum of earth beneath him was so thin that he appeared quite in the air, while he sat with his elbows on his knees, entertaining all earthly thoughts, or thoughtless, while [we] looked directly beneath him through much air to a fair and distant landscape beyond."[26] Thoreau's sense of space, often pictorial, concentrates on perspective (particularly after he had read several works by English painter William Gilpin), yet Thoreau focuses as well on unusual perspectives and multiple spaces. In the passage above he bypasses a traditionally important and vibrant symbol, the bridge, as he chooses to look underneath it. As Martin Heidegger notes (and certainly he is not alone in noting the symbolism of bridges), a bridge connects and "guides" us to a central location. "It brings stream and bank and land into each other's neighborhood. The bridge gathers the earth as landscape around the stream."[27] Thoreau stands on the sandbank at the threshold of this landscape and suspends the movement and "gathering" involved in the man's crossing. He points out that this possibly thoughtless man rides on the surface of things, ignorant of other landscapes.

At the conclusion to *Walden*, Thoreau criticizes people for not seeing deeper into their space. "Most have not delved six feet beneath the surface, nor leaped as many above it" (586). Thoreau uses a concrete image not merely to convey the metaphorical implication that people should be more introspective, he means this image literally as well. His sense of space (unfolding throughout *Walden* in consistent motifs, including passages on his house, the woods, and the pond) entails a sense of activity in multi-dimensional space. In the bridge passage above, Thoreau as perceiver stands outside multiple spaces; as an occupant of Walden Woods, he joins his experiences in a complex layering of spaces.

Thoreau's reconceptualization of nineteenth-century space encompasses the entire definition of open and enclosed space—the meaning of indoors and outdoors, boundaries, and shelter. In this reconceptualization, Thoreau examines the interwoven qualities of space—how we act in and perceive our environment and its dimensions and depths, what directions things move through spatial fields and the paths taken. Thoreau's understanding of movement as a spatial quality often guides his descriptive writing. In the portrayal of "a village of busy men" in *Walden*, Thoreau describes lines, angles, and planes. He finds village men sitting "in a row . . . their bodies inclined forward and their eyes glancing along the line this way and that . . . or else leaning against a barn" (456). He himself "never hesitated at a gap in a fence" in order to escape from the town or "launch myself into the night" (457). While other writers might see just the lazy routine and near motionlessness of these village men—suggesting the static nature of the place—Thoreau not only describes motion and suggests motion, he adds his own course of motion to the scene. The village's space is determined less by occupation than by registering lines of action such as Thoreau's path through that space.

Thoreau's concept of space as an environment encompassing various sites and lines of activity precedes Gestalt theories of field forces, a concept adopted from physics into psychology and used in behavioralist-based architectural theory in the twentieth century. A field could be represented, although inadequately, by a map of places, routes, and patterns of travel, taking such forces into account as weather, food production, consumption, travel, and so on. Jon Lang explains that "all forces, activity, and behavior occur in some field of environment. . . . The state of the field is a resultant of all the forces acting in the field."[28] This theory discounts the idea of static or empty space as a void waiting to be filled. Space cannot be considered a priori because it relies on the movement of things, bodies traveling in and through areas. Space and things in space form a reactive whole. Thoreau seemed to understand this principle intuitively.

Instead of describing his house as static space, Thoreau describes changes occurring in that field. Even when he leaves the house unoccupied, he anticipates activity. He writes that "[w]hen I return to my house I find that visitors have been there and left their cards, either a bunch of flowers, or a wreath of evergreen, or a name in pencil on a yellow walnut leaf or a chip" (425). Anticipating activity independent of his own life, Thoreau looks expectantly for changes in his environment. Significantly, instead of the two-dimensional calling cards used in society, visitors here leave three-dimensional, natural items. Even a visitor's written name takes on solidity and texture from a wood chip. Visitors mark Thoreau's environment, remaking it through the addition of objects to his space; their travel through his house creates fluid space.

Henri Lefebvre identifies the basic duality of space as consisting of both place and motion: "Is not social space always, and simultaneously, both a *field of action* (offering its extension to the deployment of projects and practical intentions) and a *basis of action* (a set of places whence energies derive and whither energies are directed)?"[29] The nature of a place determines the energies expended in that space. When Thoreau walks through snow-filled woods, he notices the architectural forms produced by snowdrifts. The "architecture of the snow" emerges from suggestions of form

WALDEN;

OR,

LIFE IN THE WOODS.

BY HENRY D. THOREAU,

AUTHOR OF "A WEEK ON THE CONCORD AND MERRIMACK RIVERS."

I do not propose to write an ode to dejection, but to brag as lustily as chanticleer in the morning, standing on his roost, if only to wake my neighbors up. — Page 92.

BOSTON:

TICKNOR AND FIELDS.

M DCCC LIV.

Title page of Henry David Thoreau, *Walden; or life in the woods,* 1854, showing Thoreau's hut at Walden Pond, Massachusetts.

"from a wall or bank" in combination with what he describes as the gaseous energies of the storm, to create snow curving "gracefully upwards into fantastic shapes."[30] Snow piled on tree limbs reminds him of "miniature Chinese walls, zigzag over hill and dale," and along his labyrinth path "singular portals and winding passages were left between pitch pines."[31] As a "basis of action," the woods, after a heavy snowfall, determine Thoreau's path. In *Walden*, Thoreau comments on the ways the seasons change his environment. When the ponds froze "they afforded not only new and shorter routes to many points, but new views from their surfaces of the familiar landscape around them" (538). The seasonal changes and the landscape affect and alter both mobility and perception.

Thoreau's field, his site at Walden, contains paths of other living things, and his house provides new spaces for them. Birds, wasps, squirrels, and hares all share his house, finding their own uses for his cabin. "The wasps came by thousands to my lodge in October, as winter quarters, and settled on my windows within and on the walls over-head" (513). When Thoreau gets up in the morning, he wakes up a rabbit under his floor, who rushes out "thump, thump, thump, striking her head against the floor timbers in her hurry" (545). Thoreau describes these creatures as sharing space, moving through his space, and forming their own spaces. Thoreau watches "while the birds sang around or flitted noiseless through the house" (411). The trajectory of birds through the house fills the seemingly empty space between the four walls.

As Thoreau himself moves, space changes around him. In *Walden* he identifies the mutability of space and incorporates this recognition into his writings. He blurs the boundaries between outside and inside, both between nature and house, world and self. He moves to Walden to create his "sedes" in the sense of creating an estate for himself and a manner of life. Playing with the idea of a "sedes" as both seat and estate, he writes that "[w]herever I sat, there I might live, and the landscape radiated from me accordingly" (387). Every new position offers new perspectives. Thoreau challenges the concept of the outdoors as a stable "place" which civilized people can move in and out of at will; he sees how each new element moving through space creates a new field of activity.

Thoreau's formation of his environs draws on his own inner activity as it is influenced by the force of the space itself. He describes his new house as "a sort of crystallization around me [which] reacted on the builder" (390). He fuses his own bodily space with his new environment. "What of architectural beauty I now see, I know has gradually grown from within outward, out of the character and necessities of the indweller and builder."[32] His body suggests the form of the building, but he also changes his expectations during the construction of his new life. He recognizes his contribution to the definition of space.

When Thoreau moves into a half-finished cabin, he brings the outdoors within, erasing the usual defined border between in and out, and blends the distinctions between the two spaces. Unprepared for winter, Thoreau records his occupation of his house in positive terms, celebrating conditions with which most people would find fault and look forward to correcting. The house lacked plaster, chimney, tight fittings, and insulation, yet he describes his house as having a "clean and airy look . . . its timbers were saturated with dew . . . it retained throughout the day more or less of this auroral character" (390). Winds blew through the chinks in his house, and he tells us "I did not need to go out of doors to take the air, for the atmosphere within had lost none of its freshness" (390). He does not sit indoors but "behind a door" (390). He lives near the rain, keeps the birds as neighbors and wakes up to hear mosquitoes near his bed (390, 393). His house takes on a porous quality and the term "indoors" loses its strict meaning.

Thoreau reverses a trend towards what Foucault calls "specified and functional" rooms and returns to the eighteenth-century idea of "undifferentiated space."[33] "Instead of no path to the front-yard gate in the Great Snow,—no gate,—no front-yard,—and no path to the civilized world!" (424). The normal functions of parts of his house—the door, walls, rooms, and yard—change or disappear entirely; his entire house "was all entry" (427). Space depends less on intended function than on usage, and he knows that a house which has the intended function of serving as a home might turn into "a workhouse, a labyrinth without a clew, a museum, an almshouse, a prison, or a splendid mausoleum instead" (345). He frees his space from these static, predetermined configurations.

The boundaries of his house are less distinct than they might appear, since, despite solid walls, roof, and floor, nature invades the house. Thoreau invites and celebrates this destruction of boundaries:

> No yard! but unfenced Nature reaching up to your very sills. A young forest growing up under your windows, and wild sumachs and blackberry vines breaking through into your cellar; sturdy pitch-pines rubbing and creaking against the shingles for want of room, their roots reaching quite under the house. (424)

Again, Thoreau's space contains directions of movement—up, under, through, and in—which create a dynamic, almost threatening, space. Although this space seems too crowded, he remarks on the following page that "[t]here is commonly sufficient space about us" (425). The desire for the fusion of natural and civilized space drives Thoreau to embrace this invasion of his house.

Thoreau also revels in turning his house inside out for regular cleanings. He finds it "pleasant" to see his furniture outside, while the sight of table and chairs in the woods provides the reader with an image of a changeable, nearly surrealistic site: "A bird sits on the next bough, life-everlasting grows under the table, and blackberry vines run round its legs; pine cones, chestnut burs, and strawberry leaves are strewn about" (412–13). With anticipation, Thoreau works towards the dissolution of all culturally fabricated boundaries.

At other times, with the transformation of boundaries, even the solid form of Thoreau's body becomes porous to outside influences. During evening walks "the whole body is one sense, and imbibes delight through every pore. I go and come with a strange liberty in Nature, a part of herself" (425). Thoreau pays attention to the influx of information from all his senses so that he registers how every corner of that space is filled. Sounds and smells travel from distances to move through his space; he identifies the presence of a traveler on the road by pipe scent, and the "whistle of the locomotive penetrates my woods summer and winter" (425, 414). Hawks and pigeons transcribe his space, circling overhead or flying through the woods (413). He notices how bodies connecting in space produce change, as when "the sedge is bending under the weight of the reed-birds flitting hither and thither" (414). The transformation of the spatial imagination does not occur through supernatural means but through Thoreau's perceptual acumen.

III

Thoreau tells us in *Walden* that he desires "to speak somewhere without bounds" without the "stone-walls piled up on our farms"

(580, 577). Throughout this work he draws what Gaston Bachelard calls a "topoanalysis" of his own space, repeatedly describing his life through the condition of his house and the relative positions of house, woods, pond, railroad, town, and even distant hills. In considering Thoreau's physical and psychological space, I am undertaking Bachelard's question of "the two-fold imaginary geometrical and physical problem of extroversion and introversion."[34] Bachelard limits his phenomenological approach to the "imaginary" and skirts this issue of the interaction between physical and mental space. Thoreau, however, immerses himself in his world until his metaphors and his experience become indistinguishable, and the boundaries between thought and experience dissolve.

Traditionally, scholars of Thoreau trace a linear progression in his work from authorial experience to written description to metaphoric meaning. In "Delving and Diving for Truth," Joseph Allen Boone cites several of the images noted in this study to support his claim that "the cumulative effect" of "references to probing earth and water . . . triggers the reader's insight into Thoreau's metaphoric quest for meaning."[35] Although Thoreau's house-building lends itself to metaphoric interpretation, his life in the woods—not the metaphors—contain the meaning that Thoreau presents to us. In addition to the correspondent truths between the material and spiritual worlds that Transcendentalists sought, Thoreau points to pervasive truths. Transcendentalism's influence on the nineteenth-century spatial imagination originates from the connection of real experience of spatial configurations with a transcendent state. The difficulty in determining Thoreau's "topoanalysis" of his own space and his understanding of physical space lies in his use of metaphors. Because metaphors rely on analogies between concrete materials and abstract concepts, they complicate any attempt to distinguish between mental and physical space. Thoreau's metaphors, however, embrace real things, emphasizing corporeal reality as much as any symbolic content. In concluding his chapter "Where I lived and what I lived for," Thoreau's language swings between real experience and idea. "Let us settle ourselves, and work and wedge our feet downward through the mud and slush of opinion, and prejudice, and tradition, and delusion . . . till we come to a hard bottom and rocks in place, which we can call *reality*" (400). This statement travels unreservedly from feet to prejudice to rocks. Although his writing sometimes leads us to believe that he does swing between two extremes—physical and mental states—this perception does not accurately describe Thoreau's understanding. Thoreau melds the physical and the mental (including the spiritual) into a "reality" which incorporates, in a manner of speaking, both body and soul. Digging one's toes into the dirt and perceiving this experience fully connects one to a "higher" state.

In 1909, Jacob von Uexkull suggested that each living being operates in its own particular world (*Umwelt*) and that these worlds overlap. Animals carry around different concepts of space and live in different spaces, all of which form part of a single environment: "the animal forms its nature according to its special needs."[36] Uexkull posits that amid this plethora of spaces may be a higher space of which humans are unaware. Uexkull, according to Stephen Kern, "speculated [that] there may also be higher worlds of greater dimensions that we are unable to see, as the amoeba is unable to see the stars in our sky."[37] Working towards this theory from the other end and half a century earlier, Thoreau understood nature as a way to explain transcendent realms, and he frequently implies that nature and transcendence conjoin. He saw that experiencing all aspects of the natural world was therefore crucial to attaining a spiritual world.

Thoreau spends several pages of the "Friday" section of *A Week on the Concord and Merrimack Rivers* considering the nature of physical space and the possible nature of transcendent space. His comments probe the question of defining the qualities and boundaries of space. He grapples with the idea of interior and exterior spaces. He presents the quandary of space in the most blatant, although not the clearest, manner found in his writings. By the time he writes *Walden*, Thoreau seems to have internalized these dilemmas, and to some extent resolved the issue of space. In *A Week*, the task of interpretation becomes complicated by Thoreau's love of paradox and inversion, yet his discussion centers around the assertion that all space—interior, exterior, spiritual and real—has physical traits. Transcendence does not involve escapism from this real world. Thoreau asks, "[a]re we to be put off and amused in this life, as it were with a mere allegory? Is not Nature, rightly read, that of which she is commonly taken to be the symbol merely?" (310). Thoreau rejects an understanding of the world that relies on metaphors. Nature does not function as a symbol of Heaven; Nature "is" Heaven. Whenever Thoreau complains that people see the "surface" only, he means that we do not use our senses to their full capacity. He explains "reality" as both a physical space and a spiritual state.

The mixture of nineteenth-century metaphysical language, mystical images, and metaphorical comparisons complicates Thoreau's discussion of space. He emulates popular religous images when he describes nature as an "echo" of a fairer realm—calling to mind the traditional image of Elysian fields or descriptions of the New World as an edenic garden—of which we "live on the outskirts" (308, 311). He exhorts his readers to develop their spirituality through the use of their spiritual eye, since the physical eye is limited and "may see for the hand, but not for the mind" (312). He describes the spiritual realm, that "OTHER WORLD," as alternately "other mansions" and an "immaterial starry system" (306, 312).

In all these statements Thoreau is still relying on metaphoric meaning to search for correspondent truths because, as he tells us, he believes in the relation of the real to the ideal. In *Walden* he emphasizes his belief in the Absolute. A mountain may have "an infinite number of profiles, though absolutely but one form. Even when cleft or bored through it is not comprehended in its entireness" (554). Thoreau searches for this "one form" through the collection of facts and observations about material form from multiple perspectives. Yet in his search he does not limit himself to this exacting separation of the Absolute from the multifarious world. He does not reject his joy in fluctuating circumstances or despair of finding the one form in many places. He celebrates multiple perspectives of "the Real" as embodying, not simply indicating, a spiritual state. "Here or nowhere is our heaven," he says in *A Week* (313, 308). To this spiritualized space he ascribes all the qualities normally associated with the material world, contrasting the "spaces of the real, the substantial, the eternal" with the "material universe" (313).

Thoreau affirms the power of the mind to discover "real" space and its qualities. "The mind knows a distance and a space of which all those sums combined do not make a unit of measure,—the interval between that which appears, and that which is" (314). Metaphorical comparisons to science—or any other field—can lead one only as far as current scientific discovery. Thoreau places his faith both in future exploration of spatial fields and in the mind's power intuitively to reach beyond scientific constraints. The mind not only understands the terms of measurements, it also sees the spaces in between. Thoreau takes a mystical statement, an affirmation of unusual and inexplicable powers, and actualizes it as a spatial phenomena. His "interval" of space has dimensions, contains directional movement, and simply demands that we fine-tune our senses since this spiritualized space "is not far off geographically" (306).

Thoreau maps his topography for the reader in an unsystematic way but with a consistent understanding of space as containing many overlapping and interactive spaces. Transcendence does not necessarily mean shedding past conditions like old skin. "In space, what came earlier continues to underpin what follows."[38] For Thoreau, changing states is a matter of weaving more intense connections with one's present state. In *Walden* he surmises that music can transport a working man out of his preoccupations. Music tells the farmer that "stars twinkle over other fields than these." The farmer's resultant desire to change his condition is formulated in a plan "to practice some new austerity, to let his mind descend into his own body and redeem it, and treat himself with ever increasing respect" (500). The farmer, instead of moving, or becoming more educated or more wealthy (or less wealthy), stays rooted in the same place and allows that place to change around him as he modifies his perspective. In this passage the movement outward to the stars turns smoothly into a descent inward.

Rivaling Emerson's cosmography, Thoreau's space emphasizes movement within both the self and the physical field containing one's body. Transcending to "another and purer realm" means looking outward while reaching inward (309): "It is easier to discover another such a new world as Columbus did, than to go within one fold of this which we appear to know so well; the land is lost sight of, the compass varies, and mankind mutiny . . . (310–11). Although transcendence was and remains associated with movement up, above, and beyond, for Thoreau it involves moving inward as well as outward and incorporates an intensity of personal effort.

Emersonian space expands outward—"The life of man is a self-evolving circle, which, from a ring imperceptibly small, rushes on all sides outwards to new and larger circles, and that without end" (169). In contrast, Thoreauvian space moves and expands *inward*. "This world has many rings, like Saturn, and we live now on the outmost of them all. . . . What do botanists know? Our lives should go between the lichen and the bark," he declares in *A Week* (312). Thoreau works with the idea of multiple spaces years before such theorists as Uexkull; his sense of space takes him between lichen and bark, between material objects that we can identify into the spaces that they create. He looks for "gaps" in the fence, new and unpredictable spaces that afford new opportunities to pass through old boundaries. "[W]e think that if rail-fences are pulled down, and stone-walls piled up on our farms, bounds are henceforth set to our lives and our fates decided. . . . The universe is wider than our views of it" (577).

After Einstein's theory of relativity, which rejected the "vague" idea of space and replaced it with the concept of "motion relative to a practically rigid body of reference," "the number of spaces increased beyond calculation."[39] Half a century before Einstein, Thoreau theorized that we have the power to investigate these multiple spaces. "Surely, we are provided with senses as well fitted to penetrate the spaces of the real, the substantial, the eternal, as these outward are to penetrate the material universe" (313). As of now, he says, we "live on the outskirts of that region" (312), the spaces within/without us. Thoreau takes Emerson's concept of the world as concentric circles expanding towards the Oversoul and turns it inside out. For both though, the key to transcendence lies in individual power. Thoreau's understanding of the physical qualities of space does not rely on materialism in the traditional, Kantian sense. Unlike this century's critic, Bachelard, who bypasses the problems arising from privileging space over imagination, Thoreau describes physical and mental worlds as cooperative. Bachelard understands space as created by the individual consciousness: "Space calls for action, and before action, the imagination is at work."[40] Thoreau's answer is that his reader "needs not only to be spiritualized, but *naturalized*, on the soil of earth" (307). His philosophy of life calls for the simultaneous growth of both spiritual and physical understanding.

The American Transcendentalists including Thoreau did not accept transcendence as the end of stability but as an ongoing dynamic process. In a sense they often emphasized transcending over the transcendental, the dynamic verb over the static noun. Even the style of many Transcendentalists, according to Catherine Albanese, was a style of "kinetic motion."[41] Transcendentalism has become an overused term; even from the beginnings of its use as a philosophical term, by the time it traveled from Germany to England (through Kant's writings) and then to the United States (via Coleridge), transcendence had more than one specific meaning. References to Transcendentalism, transcendence, and transcendent things occur with high frequency in print; even twentieth-century Marxist critic Henri Lefebvre mentions transcendence periodically. In general—and the following is oversimplified—two categories of thought exist for things of a transcendental nature; one group emphasizes the completion of experience (things which are transcendental) or objects (a transcendent God). The other group emphasizes the process—transcendence is something one does, transcending out of one place, condition, or state into another.

Lefebvre speaks about the natural ability of living bodies to create space through the example of a spider that "occupies a space which it engenders according to its own lights: the space of its web, of its stratagems, of its needs."[42] One cannot separate the spider from its space, since it "is already capable, just like human groups, of demarcating space and orienting itself on the basis of angles. . . . The spider has a sense of right and left, of high and low. Its 'here and now' (in Hegel's sense) transcends the realm of 'thingness', for it embraces relationships and movements."[43] The spider exemplifies the way living beings create their own space, and the way this space constitutes a field of action, with dimensions, directional movement, and activity within that field. The spider's space "transcends" older concepts of material objectivity, because this type of space incorporates movement; the spider operates on a space-time continuum. For Lefebvre, transcendence means moving from one level of understanding to another; it also implies a shift to a more complex comprehension.

Thoreau's idea of transcendence may not be too dissimilar from the secular critic Lefebvre. Thoreau calls for greater perception and understanding in which a person transcends to a finer understanding both of spiritual and physical being. "And we are enabled to apprehend at all what is sublime and noble only by the perpetual instilling and drenching of the reality which surrounds us" (399). Given Thoreau's understanding of "reality" as *more real* than the material world as commonly understood, his assertion that "we crave only reality" speaks to the mystical tradition of sight as insight.

3

Dynamic Space: *The Enormous Room*

I

In CUMMINGS' SEMI-AUTOBIOGRAPHICAL NOVEL *THE ENORMOUS ROOM,* C and B (characters based on Cummings himself and his friend William Slater Brown) volunteer for the Norton-Harjes Ambulance Corps at the start of World War I. They quickly earn a poor reputation with the American military command for fraternizing with the French staff, or as the American military called them, the "dirty Frenchmen."[1] Having already mastered French slang during their stopover in Paris, C and B converse with the French cooks and mechanics in their unit and write enthusiastic letters home about the French in preference to the Americans. In his letters, B's humanitarianism occasionally spills over to favorable comments about the German enemy. When the French government intercepts letters written by "B," they question both men and then send them to La Ferté-Macé, a detention center.

C breaks with traditional, nationalistic wartime views to act on his own, prizing independence and uniqueness. C passes up an opportunity to deny involvement with his friend B's letter-writing, remaining loyal to him and following him to detention. C transforms this detention site into "The Enormous Room," a heterotopia of transcendence. He inverts an already inverted and differentiated space (church buildings converted for use in the war) and records his exploration and creation of the interrelated elements peculiar to this site.

The transformation of La Ferté-Macé into the Enormous Room is not strictly an act of the imagination. Intending La Ferté-Macé to function as a heterotopia of deviance, the French government began the process by converting church buildings into a detention center. The metamorphosis of the buildings from one heterotopic use to another added dynamic qualities to the site. As a place which could "reflect and speak about" other places, according to Foucault's definition of a heterotopia, this particular site had developed the character of a palimpsest even before C's arrival.[2] C immediately recognizes this layered quality and describes the place in terms of multiple places: in his observations the room functions as a church building, a classroom, a playground, a cultural way station, a political rallying ground, and a home.

Just as a theater, in Foucault's example, contains a rectangular stage where other "places" are evoked, the functions of several spaces overlap and combine in the Enormous Room. In discussing these various spaces in one, it would be a mistake to limit the discussion simply to function. Henri Lefebvre delineates the three critical aspects of space as structure, function, and form.[3] In *The Enormous Room* the physical aspects of structure interact with the room's intended and reconfigured functions, maintained or changed through communication (form); the interaction of all these aspects creates dynamic space.

In the detention center various structural, functional, and formal factors—external and internal—create a space where sites of spiritual nourishment, education, self-expression and cultural exchange overlap. External factors include the design and materials of the building, and the use to which the building is put, and these factors play a major role in stimulating the characters' perception and imagination. Because the prisoners are contained in one large room rather than isolated in separate cells (thirty to forty men occupy this one room), personal interactions increase at exponential rates.

The architectural features of the room—most notably the high ceiling and "ecclesiastical windows"—constantly remind the occupants of its original function as a church. These windows endow a place meant to confine and subdue with spiritual, transcendent qualities. The religio-mythic element in the narrator's perception overpowers the room's current intended function. C describes his detainment in terms of a pilgrimage and recognizes other characters and places as "signposts" on this pilgrimage; the supposed "deviation" of these prisoners originates in spiritual enlightenment.

Cummings interweaves the structural elements of La Ferté-Macé with references to John Bunyan's *The Pilgrim's Progress*—layering each formal reference onto the place's intended function.

For example, when C names four men as the "delectable mountains," he interjects an image of a physically large geographical formation onto a person and into a room. He also adds allegorical significance to the place by replicating one feature of Christian's pilgrimage. (In another inversion by Cummings, C's 'journey' takes place in dreamlike reality, whereas Christian's journey entails a realistic dream; "place" takes on multiple meanings.) Cummings transposes Bunyan's cognitive, imaginary map onto real places, beginning with the "Slough of Despond" (the ambulance station) and ending with the "Celestial City" (Paris). C's release and subsequent trip to Paris completes the transformation of mental geography into physically real, privileged sites.

As the characters move between sites they create a set of relationships between these places. If a site consists of "relations of proximity between points or elements," then space, as the "relations of sites," recalls the concept of Gestalt psychology's "field force."[4] The field force concept, adapted first from physics and then reapplied by architects like Jon Lang, suggests ways in which elements interact within a given environment: "All forces, activity, and behavior occur in some field of environment. Forces in the visual field, as in mechanics, have an area of application, a direction, and a magnitude. The state of the field is a resultant of all the forces acting in the field."[5] As Lang implies, in a field the site includes and takes into account various forces (such as weather, electricity, biological growth), activities (movement, food consumption, consumerism), and behaviors (displays of emotion, communication). The field concept in architecture utilizes behavioralism and Gestalt theory to determine how we understand our environment. Lang emphasizes the primacy of sight in determining a complex environment, arguing that a person's movement in space "involves transformations in the optic array."[6] Accordingly, architectural theory should take into account all types of surfaces including opaque planes (those parts of a building which hide another part of the structure), surface curves, textures, and contours.

Other architects expand Gestalt theory into a "Body-Image Theory" and point out other sensory systems besides sight which influence our perceptions such as auditory systems, postural orientations, and haptic systems.[7] Due to the interactions of our multiple sensory systems with an environment that changes even with the varying amounts of light during the day, the function of a place should not be considered a static element. The "forces" in a place constantly interact and create dynamic situations. Cummings was fascinated with dynamic, or in contemporary terms "kinetic," effects; he looked for kinetic effects in diverse environments.

Because he is attuned to his environment, C refers frequently to the noise and activity in the room. The dimensions and shape of the room decrease privacy and increase the volume of noise so that solitude is difficult, and the activities of others easily attract attention. These factors both degrade and uplift the men. When a new (and pompous) detainee lights candles at night and carries on conversations, he disturbs the sleep of all the other men (142). C relates his contradictory feelings about the effect of the room on the men; its size and openness in conjunction with its imprisoning function encourages dehumanization. The room, therefore, has sinister qualities. Mealtimes occasion a transformation of the men into a mob, "a fluent and numerous cluster of vital inhumanity" (66). Bedtimes occasion another type of loss of humanity in sleep; as men begin to snore, a "candle flickered a little as if darkness and itself were struggling to the death" (81).

However, because of the room's large dimensions, the height and breadth of its walls, and its shape (rectangular, with no nooks or corners), a large number of prisoners can interact as a group and give way to free self-expression and rebellion. In response to the pompous man and his candle, the rest of the occupants assault him with animal noises, "neighings he-hawings roarings bleatings growlings quackings peepings screamings bellowings" (143). Although acting as part of a group, each man contributes an individualized, creative noise. C remarks that "the racket [. . .] set The Enormous Room suddenly and entirely alive.Never have I imagined such a ménagerie as had magically instated itself within the erstwhile soggy and dismal four walls of our chambre."[8] The men completely change the room's function; instead of a place of confinement, restraint, and figurative death, the "insurgents" create a playground, an imaginatively "alive" place (143). The resulting activity and reverberating noise convert the room into a dynamic space. To call to mind an opposite situation, one has only to compare this description with a monk's cell—or a prison cell—with its small dimensions, barrenness, and single occupant following strict rules.

The narrator's perceptions of this place stimulate his imagination; the room does not simply *function* as a place for dynamic activity—it is *physically* transformed into a dynamic and mobile place as structure and function interact. In terms of its construction the room is "in shape oblong,about 80 feet by 40,unmistakably ecclesiastical in feeling" (50) with pillars and cathedral windows. Cummings' initial perception, however, imparts anthropomorphic life to the room. C arrives at night, and when he enters the place "the hitherto empty and minute room became suddenly enormous;weird cries,oaths,laughter,pulling it sideways and backward,extending it to inconceivable depth and width,telescoping it to frightful nearness. From all directions,by at least thirty voices in eleven languages [. . .] I was ferociously bombarded" (43). The space expands and fluctuates; the punctuation and pace of his phrasing indicate the unstable motions evoked by the room. The interconnections that the reader makes in this compressed passage (in which the alliterative phrases "sideways and backwards" and "depth and width" also mirror each other in meaning) function much like Gertrude Stein's description of the room in *Tender Buttons*, a work which Cummings admired and quoted during his 1915 Harvard commencement speech "The New Art."[9]

In this address, Cummings discusses various modern movements in art, music, and literature. His essay attempts to connect the "new art" to recent advances in realism; in discussing Stein's "great imagination" he points to the "traces of realism, similar to those which made the 'Nude Descending the Staircase' so baffling." While Cummings does not elaborate on this connection between Duchamp's style and that of Gertrude Stein, he had noted at the start of his speech that, in Cubism, "by using an edge in place of a curve a unique tactual value is obtained."[10] The assumption that every generation makes about realism—that realistic artwork can be easily understood because it is supposedly real—leads to frustration. Artists render "the real" through different conventions; the multiple brushstrokes in Duchamp's "Nude" suggest the movement of legs down a stairs, a realistic rendition of motion through new techniques of representation.

Cummings mentions that Stein cares for "the beauty of words themselves" and her preference for the reality of words themselves over the meaning of syntactical groups results in her particular style:

A sound.
Elephants beaten with candy and little pops and chews
[. . .]
Please pale hot, please cover rose, please acre in the red [. . .][11]

As with Duchamp and the movement in "Nude Descending the Staircase," Stein attends to the tactile sensations of words. Cummings, describing C walking into the Enormous Room, renders another realistic phenomena—the effect which sound has on spatial perceptions. "[W]eird cries,oaths,laughter,[pull the room] sideways and backward,extending it to inconceivable depth and width,telescoping it to frightful nearness." Just as Cummings noted the inherent "logic" in Stein's approach to words, he logically—literally—explains a particular phenomenon of perception in which the perceiver becomes disoriented by the growing volume of uncontextualized sounds. Because he can't identify the meaning of these cries and oaths, C relies on the room itself to provide context; for him, the inhabitants have transformed the room into a language field.

The people in the Enormous Room embody the essence of a mobile, multidimensional space; through them the room transcends the limitations of its walls and seems to live and breathe. The experience of walking into that room has more import than its dimensions on paper, and a crucial aspect of the room is perceptual and interactive. As the room configures itself around a living body of men, that body creates the space.

Those men who are unable to live freely through their perceptions, who cannot enjoy a unique experience of space, remain trapped in one-dimensional space. The Machine-Fixer had "l'âme sensible" (a sensitive soul) and felt acutely not only his suffering but also every detainee's suffering to the point of stagnation. For him the Enormous Room was simply "La Misère" and his "abrupt conscientious relentless difficult self lived always in a single dimension—the somewhat beautiful dimension of Sorrow" (100). His perceptions, experiences, and interior life revolve around a singular negation, his lack of liberty. The narrator, however, enjoys all the dimensional experiences offered in the room, which he describes as "like a vast grey box in which are laid helter-skelter a great many toys each of which is itself completely significant" (82). The childish government has strewn men into their toy box; C carefully removes each one for us.

C's introduction to the Enormous Room also initiates him to the room's function as a cultural meeting-place, which he describes in terms of another series of dynamic interconnections. The eleven languages with which he is "ferociously bombarded" on his first night belong to ethnic groups whose representatives have been transported from various places into one place. The narrator's perception of this room as a way station for various cultural representatives enhances the feeling of exploration and discovery; he travels across places regardless of time without physically moving his body. Links between places develop as the narrator lives with people of other nationalities and learns to tolerate their mannerisms. He and B collect a group of friends, among them a Gypsy (Josef Demetre, the Wanderer), a black man from Paris (Jean Le Nègre), two Poles (Surplice and Zulu), a Mexican (Philippe Burgos, or Mexique) and an African (Afrique). They share language lessons, stories of their homes and families, and other reminiscences.

The narrator, however, is hardly a model of enlightened tolerance. C defends other detainees while simultaneously reveling in stereotypical traits.[12] The Wanderer invokes the "sheer kingliness of poverty" (162) in an image often associated with Gypsies, and Jean Le Nègre, whom C defends against prejudiced guards, is celebrated by C as "Boy, Kid, Nigger with the strutting muscles" (215). In the latter, Cummings deftly includes three derogatory and belittling stereotypes in one short phrase. Christopher Mott remarks that in the character of Jean, "primitivist assumptions manifest themselves in a romantic racism" that praises innocence, childishness, and physical traits.[13] It appears thoughtless (if not downright obnoxious) to current readers for a well-educated, privileged member of the majority to praise people consistently in terms of stereotypes. Yet in defense of the narrator, it can be said that in other ways he "encourages our sense that Jean is a human being"[14] and is acutely aware of his situation. C also relates how Jean Le Nègre flouts his muscular build and uncivilized behaviors to intimidate and control the guards, as one of several ways that his fellow detainees use stereotypes as tools of power against authority.

Through proximity with various men and by means of conversation, C goes "around the world a number of times" and learns about places he'd never seen (79). His particular group of friends present "le nouveau" (the new arrival) with interesting information about their "lives habits and likes" (79). Later, C describes other physical and imaginative attributes which they possess and which convey these other worlds to him. The Wanderer, who appears Arabic, boasts a beard "blacker than Africa. [. . .] Than imagination" (162), and when he leaves for prison, he takes with him "unspeakable sunlight,and the dark keen bright strength of the earth" (167). By collecting images of external things like sunlight and earth, and valuing the inner worlds of the detainees, C progresses on his pilgrimage and develops his own sense of spiritual freedom.

Part of C's self-education involves refashioning social and political realities to fit his personal conceptions. The room and its occupants create unanticipated areas of communication, including the exchange of spiritual ideas, cultural values, and concepts of self-government. The room acts as a classroom, a center of rebellion, and an embryonic utopian community. Like other heterotopias, the Enormous Room serves to "suspect, neutralize, [and] invert" the rest of society. C discovers and nurtures the site's potential to "designate, mirror, reflect," the system of detention and imprisonment with which the society supports the military's activities.[15]

C connects the theme of spiritual freedom to political and social freedom, associating individualism and personal autonomy with both personal growth and human rights. To the French government, the detainees are transgressors who have abused their freedoms and who continue to transgress—criticizing the military, uprooting and searching for families from whom they have become separated without regard to state boundaries, disobeying wartime and peacetime rules and regulations. The Camp de Triage at La Ferté-Macé serves to separate these men from the rest of society in order to limit the contamination of ideas and actions; ironically, for those people within, the room serves as a classroom, a place of concentrated ideas available to be learned. The men discuss abstract concepts such as liberty and human rights in connection with the conditions of their detention: the unfairness of the committee members who decide which detainee should be sent to prison; the poor quality of the food; the unsanitary condition of their room (a place combining sleeping facilities, open toilets, and damp, ill-heated air); and the tragic separation of men from their families.

Cummings contrasts the conditions of the detention center and its occupants with society, particularly the power structures of the military, the government, and the bourgeois citizenry. C criticizes the government for its lack of insight into individual character and its enjoyment of punishing people. The mixture of criminals with ordinary people in the detention center proves volatile (particularly after the arrival of The Fighting Sheeney), and C wonders why a delicate man like the Wanderer has been detained. In rhetoric that mocks the lack of insight of "great" men, C asks, "O little,very little,government français,and you the great and comfortable messieurs of the world,tell me why you have put a Gypsy who dresses like Tomorrow among the squabbling pimps and thieves of yesterday . . . (165). With condensed punctuation and compressed spacing, Cummings emphasizes the smallness of the government in comparison to the Gypsy, whom he designates with a capitalized and progressive "Tomorrow." Through periodic phrasing, Cummings compares the "great and comfortable messieurs of the world" with the "pimps and thieves of yesterday." His project of syntactic innovation reflects—on the level of the space of the page—the reformation of meaning he applies to spaces of human occupation. On both levels, new paths and new connections are etched into traditional spaces. A "prison" is now a pathway to a spiritual and social goal—freedom. C demonstrates how the place of the Enormous Room reflects back and critiques society for its lack of values (fulfilling the purpose of a heterotopia) and how it provides a microcosm of a utopian society by promoting freedom, friendship, creativity, and spiritual enlightenment.

The spiritual import of a place depends on individual perception. Other heterotopias work through often unspoken, cultural conventions (we buy a ticket to the theater and understand that we must participate in a suspension of disbelief), but a transcendent space is necessarily an individualized creation.

As a heterotopia of transcendence, the Enormous Room encourages the activity of individual life, suggesting how to go beyond the power structures and behavioral regulations of society to experience ephemeral realities. C transcends the intended limitations of place to participate in interior life through exterior reality—or in Emersonian terms, his "transparent eyeball" sees the presence of the Oversoul in nature (24). This dynamic activity occurs not simply through a static triad of structure, function, and form but in an active *field* of forces, activities, and behaviors, where structure and function interrelate.

When one site counters and critiques another, binary terms proliferate—inversion, reflection, mirroring, opposition. Lefebvre warns against structuralism's positivistic system of opposites, praising instead the flexibility of the triad.[16] The triad (conceived here as structure / form / function and force / activity / behavior) serves as a basis for conceiving interrelated factors and building a diagrammatic grid or web of relationships. From the triad it is much easier to add in other interconnections, and so approach an understanding of connections within a field rather than beginning—and ending—with binary opposites. While an exploration of opposites stimulates critical thinking, it may not necessarily lead into the labyrinth where each choice branches out progressively.

Cummings appears to favor dualisms for the depth of meaning brought by the contrast of opposites. As a detainee, C feels alive and free—his previous "life" outside detention was actually "death" because of the limitations placed on his thoughts and behaviors. Detention also creates other dualities: inmates prefer dirtiness to cleanliness, and silence and noise occur in extreme. C compares the dualism of mutability and stasis to time and timelessness, and contrasts verbal and nonverbal communication. The meanings seemingly implied by dualities, however, become refracted and dispersed through the dynamics of the room. To C, "what made death worth living and life worth dying at La Ferté-Macé was the kinetic aspect of that institution" where "simple existence was happily complicated [. . .] by a fortunate violence" of new arrivals (107). Each new arrival, and the resulting interactions, creates a more complex web, one that stimulates the interest of C and B.

Simple binary opposites unravel in the Enormous Room. Instead of mirroring and *reflecting* back on society, concepts are *refracted* and dispersed. This displacement of binary thought acts as another blow to traditional space. As mentioned earlier, the word "liberty" acts as a node connecting various concepts, including both halves of the dirtiness / cleanliness dichotomy. In discussing the purity and perfection of this abstract concept, C describes "the unspeakable cleanness of Liberty" (101). Dirt represents a different kind of liberty, the freedom to rebel against imprisonment, as C "wallowed in a perfect luxury of dirt [. . .] protesting in a(to me)very satisfactory way against all that was neat and tidy and bigoted and solemn and founded upon the anguish of my fine friends" (236). The term "liberty" splits into a double refraction (I am borrowing from and reapplying scientific language to a new context). By substituting more complex ways of perception for binary thought, Cummings is establishing his claim that space acts in dynamic ways; the figure of a web allows for more interaction than a simple path from 'here' to 'there.'

The binary group of time / timelessness might at first seem to be linked with mutability / stasis, since imprisonment implies a timeless, static, and repetitive existence. C lets his reader know, however, that the experience of imprisonment involves an understanding of time as nonlinear. He remarks that jails create an "unchanging temporal dimension" (especially for a prisoner with an indefinite sentence) which destroy traditional concepts of past and future: "how,in such a case,could events occur and be remembered otherwise than as individualities distinct from Time itself? Or,since one day and the next are the same to such a prisoner,-where does Time come in at all? [. . .] events can no longer succeed each other" (82). In the entire sequence of C's detention (narrated in chapters 5 through 12), C asserts that in prison, time consists only of "the actual Present" (83). He dismisses calendar time by declaring the preeminence of the present, creating a type of timelessness by denying the cyclic progression of time from past to present to future. Foucault remarks that a "heterotopia begins to function at full capacity when men arrive at a sort of absolute break with their traditional time."[17] The idea of time becomes refracted and dispersed in the Enormous Room, so that C describes actions taking place in sequential time, yet his sense of time becomes more spatial, more dimensional and less narrative during his detention. Characters appear and reappear, and the description of isolated events and singular impressions becomes more important to C than maintaining narrative control.

The dualism of silence / noise also begins as a binary inversion but evolves into an expression of the multidimensional quality of space. Silence, an aural "space" not controlled by words, functions as a central topoi in *The Enormous Room* and as a means of identifying the essence of humanity. Numerous illiterate, shy, or quiet characters (such as the Silent Man or the man in the Orange Cap) represent inarticulate people who possess "brilliant quietness" (71). C describes the "linguistic awkwardness" of Jan, a Danish man who gives "the impression that to give or receive an idea entailed a tremendous effort of the intelligence" (71). The rhetoric of education and ideology has not tainted the uneducated class, in C's romanticized vision. In characterizing the Delectable Mountains, (the Wanderer, the Zulu, and Surplice) and also Jean (all of whom are either illiterate or untaught in foreign languages), C emphasizes their ability to surpass their educational level in part because they have the ability to silently communicate important, spiritual truths. They epitomize physical and spiritual fluidity.

The Wanderer (a Gypsy) is an archetypal figure whose exterior form communicates spiritual qualities. C focuses on his face as a mirror of his inner state, a face "finely formed and almost fluent contain[ing] a beauty and dignity which,as I first saw it,annihilated the surrounding tumult without an effort" (73). The Wanderer, by his sheer existence, breaks down the barriers between self and the exterior. The Zulu also has "a face at once fluent and angular,expressionless and sensitive" (183). More so than the Wanderer, the Zulu poses the dilemma for C of describing the unknowable. "He could not,of course,write any language whatever . . . the secret of his means of complete and unutterable communication lay in that very essence which I have only defined as an IS" (173–74). The Zulu imparts meaning through silent emission, a transcendent communication of human essence.

Surplice and Jean Le Nègre express themselves through inarticulate sounds and represent the reverse of the Wanderer and the Zulu's passivity and silence. They actively pursue dramatic roles as the fool and the clown respectively in order to disguise their inner natures. Instead of silent radiation, both characters emit their unknowable essence through pure noise.

The personality of Jean Le Nègre, the last Delectable Mountain (a position which marks him as the spiritual epitome of this group of characters) resides in the necessary expression of his essential self, "the bright child of [his] mind" (214). Jean's dramatic dialogues in French demand all of Cummings' stylistic flexibility and frequently rely on capitalization and exclamation marks. C remarks: "il est Fort! [strong] M'sieu Jean,c'est un GEANT" [he is a giant] (210). Jean fills the entire space of the room (and the sentence) with his presence. Like his body, Jean's voice inspires C through its physicality: "His use of language was sometimes exalted fibbing,sometimes the purely picturesque. He courted above all the sounds of words,more or less disdaining their meaning" (199). In Cummings' restructuring of reality into a site of transcendence, Jean represents complete freedom—pure sound as an expression of self not restricted by any social or linguistic system. He is a mobile figure.

In asserting the centrality of dynamism in his characterizations, Cummings was acting on his reading of Futurist manifestos, which he initially encountered in Arthur Eddy's 1914 summary of contemporary art movements entitled *Cubists and Post-Impressionism*. Eddy quotes Filippo Tommaso Marinetti's "Technical Manifesto of Futurist Literature" at length, particularly Marinetti's suggested application of dynamism to literature (by expressing motion inherent in objects). Eddy also reprints sections of the essay "The Exhibitors to the Public 1912," printed in the catalog to a London Futurist show. In these selections, Umberto Boccioni and other artists assert that "what must be rendered is the dynamic sensation, that is to say, the particular rhythm of each object, its inclination, its movement, or, more exactly, its interior force."[18]

The Futurists wanted not only to describe motion but also to express the interior and abstract qualities of things: "In order that the crowd may enjoy our marvellous spiritual world, of which it is ignorant, we give it the material sensation of that world," writes Boccioni.[19] The forces active in a physical site, to apply more current critical terms to the Futurists, include interior forces.

In *i six nonlectures*, Cummings retells his time in Paris as a transcendental experience: "I participated in an actual marriage of material with immaterial things [. . .] I celebrated an immediate reconciling of spirit and flesh."[20] Cummings parted with the Futurists in the techniques he used to express interior states, yet he remained interested in the movement of objects through space. Cummings found an outlet for his interest through his conceptualization of the interior dynamics of La Ferté-Macé.

II

In *The Enormous Room*, the use of heterotopic language helps convey the fluidity of interrelated structures and events in the site—communicating this perception requires a heightened awareness of language as a dynamic site. Through the use of language, the complexities of C's experience can be conveyed including the new perspective necessary in the critique of established social spaces. Paralleling C's transformation of the Enormous Room into a heterotopia of transcendence, Cummings opens up the dimensionality of language. C recognizes communication as a critical aspect of his experience of the room; how can we perceive the dynamic qualities of physical containers of space (made of dense materials like concrete and wood) if we can only express those perceptions in static language? To match expression with perception, Cummings reworks linguistic space through innovations in style. He contracts the sentence by omitting spaces after commas, semicolons and colons; he expands the sentence vertically through the use of dashes and ellipses, and horizontally through capitalization. The ellipses also provide an escape from linear time—the narrator's thought (expressed) continues (unexpressed) eternally, despite the succeeding sentences. Cummings' prose mimics the surrealistic movements of the room as perceived by the narrator.

Cummings recalls the fluid motion of these seemingly static things (the room and the sentence) in his characters as well. The interplay between linguistic space and physical space occurs most distinctly in Cummings' invocation of the Zulu:

> He did not come and he did not go. He drifted.
> His angular anatomy expended and collected itself with an effortless spontaneity [. . .] There are certain things in which one is unable to believe for the simple reason that he never ceases to feel them. Things of this sort—things which are always inside of us and in fact are us and which consequently will not be pushed off or away where we can begin thinking about them—are no longer things; they, and the us which they are,equals A Verb; an IS. The Zulu,then,I must perforce call an IS. (168)

Cummings equates a lexical unit, the verb "is" (significantly capitalized), with physical and spiritual human existence. The Zulu exists as both a mobile personality and a verb of being. When Cummings insists that the Zulu is "entirely indescribable," he points to the elusiveness of both human nature and the odd linguistic moment of being. When used in a sentence, the verb "to be" creates a paradox of movement and stasis; as a static equation, [something is something], "to be" describes both mutable and immutable conditions.

Among a variety of methods to create movement in his poetry, many of which Milton Cohen examines in *PoetandPainter*, Cummings considers how words in context move. He theorizes that a word has both a single meaning and "momentary or transient meanings" and that each word has both "a recognizable silhouette" with "edges" and with the potential to "fuse" with other words and "melt into a movement."[21] His description of the potential for movement inherent in words sounds remarkably similar to Marinetti's ideas about context and analogy in the "Technical Manifesto of Futurist Literature."

In his personal notes, Cummings carefully copied Marinetti's call to abolish adjectives, adverbs, and punctuation. Yet he modifies these dictates to suit his needs, concentrating in his poetry and prose on all verb forms (not just infinitives) to emphasize movement. Rather than concentrate on matter, as Marinetti does, Cummings focuses on making; as he says in the foreword to *Is 5*, a poet "is somebody to whom things made matter very little—somebody who is obsessed by Making."[22]

In the "Technical Manifesto," Marinetti asserts that "to render the successive motions of an object, one must render the chain of analogies that it evokes, each condensed and concentrated into one essential word."[23] (86). Marinetti cites several passages from his futurist novel *Mafarka the Futurist*. He explains that chains of analogies—which he also terms "nets of metaphor"—are a step toward "words-in-freedom," a style of writing in which the artist "must renounce being understood." For now, Marinetti writes, Futurist poets must "lighten" nets of metaphor "too disgracefully weighed down by the plumb line of logic." By creating these nets, Marinetti hopes to both lighten the word and connect it to universal elements, thereby condensing its meaning. According to Marinetti an analogy is "nothing more than the deep love that assembles distant, seemingly diverse and hostile things"; and the connection of images separated by mental distances creates speed. "Just as aerial speed has multiplied our knowledge of the world," Marinetti writes, "the perception of analogy becomes ever more natural for man." Unlike traditional metaphors, these nets bring disparate images together, requiring the reader to make mental connections. Marinetti equates the speed of mental thought processes with the power of a dynamo.[24]

Although Cummings does not describe his ideas of motion in terms of modern engines, he comments on the movement created by combining words. To him, a word has transient meanings "whenever modified by its fellow chameleons of a context; whenever it combines with another word or words to constitute a going-somewhere, an en-routeness, a self-directing entirety."[25] He tells us that he wants to create movement through words, but how do we experience this dynamism? How do words with "edges" and single meanings reveal multiple meanings by fusing with other words in context? Words occupy space, as objects in space, and he moves words, changes the spaces they occupy, and creates "movement" similar to the speed of Marinetti's analogies. To achieve kinesis or spatial movement, Cummings explodes and remolds words by restructuring words and their phonemes; his methods include the compression, elongation, repetition, and recombination of words and phonemes, and juxtapositions and recombinations of words into startling phrases.

In order to attempt to communicate the effect of the Zulu, who is "the most completely or entirely indescribable" occupant of the room, Cummings draws on dynamic phrasing (168). C remarks on Zulu's powerful, seemingly supernatural, presence: "And always [. . .] The Zulu was beside us,or wrapped around a tree in the cour,or melting in a post after tapping Mexique [. . .] limbs' tin grace,wooden wink,shoulderless unhurried body,velocity of a grasshopper,soul up under his arm-pits,mysteriously falling over the ownness of two feet,floating fish of his slimness half a bird . . ." (184). In the poetic passages concluding C's portrait of the "Zoo-loo," Cummings draws, connects, and creates movement on multiple levels, manipulating various aspects of language: visual, phonological, lexical, syntactic, and semantic.

Cummings uses meter, rhyme, and alliteration to draw the reader into the rhythm of his description. Half-rhymes like "limb / tin" and "fish / his" predominate and blend with chiasmic, alliterative phrases like "falling over the ownness of two feet,floating fish of his slimness" with its complex pattern of "f" "o" "l" "s" and "h." The metrical pattern combined with an inventive syntax creates a cumulative effect in the second paragraph quoted. Monosyllabic words give way to polysyllabic words chained together with unspaced commas. The poetic "lines" (phrases divided by commas) move from monometer to hexameter and the metrical "feet" progress through spondee (limb's tin) to trochee and dactyl ("wooden" and "shoulderless") and include some unusual substitutions with "velocity" (a paeon) and "up under" (antibasshius). The rhythm serves to tie together this cubist / futurist piece of prose and create the feeling of "going-somewhere."

Cummings relies on context to convey his meaning. In considering the cumulative effect of poetic sequences, Roman Jakobsen cites close proximity of language elements as an important feature: "[I]n poetry not only the phonological sequence but in the same way any sequence of semantic units strives to build an equation. Similarity superimposed on contiguity imparts to poetry its throughgoing symbolic, multiplex, polysemantic essence. . . ."[26] Cummings uses similarity (such as alliterative phonemes) and contiguity (proximity of those elements) to involve the reader in concurrent experiences of a palimpsest—a textual space with width and depth. Cummings moves the reader's eye and mind around the text and into the text. To use a current metaphor, we experience the text as a hypertext; we mentally "key" on the phrase "half a bird" and may recollect the mythic Icarus or Philomel and connect their tragedies to the Zulu.

Cummings' understanding of context echoes the Futurists' idea of environment and anticipates Foucault's conception of a site. Umberto Boccioni explains in his manifesto "Plastic Dynamism 1913" that "plastic dynamism is the simultaneous action of the motion characteristic of an object . . . mixed with the transformation which the object undergoes in relation to its mobile and immobile environment."[27] If we consider the word as object, Boccioni's "environment" is equivalent to Cummings' "context." Boccioni describes a "plastic whole" as "object + environment" and calls for the artistic rendering of objects through "interpenetration, simultaneity and fusion."[28] Concern for wholeness and fusion are foremost in Cummings' work as well, in poetry that he hopes will present itself as a "heap of strains, of stresses" resolved into a "bungled breathless whole."[29] Cummings methodically molds, divides, and fuses words together in order to convey the inner plasticity of things.

The dynamic field of architectural and linguistic spaces in *The Enormous Room* intertwine to such an extent that by the end of the narrative, Cummings has taught the reader his private "code." When C is told that he will be free to leave and will be sent to Paris, he grapples with the idea. "What is Paris—a place, a somewhere, a city, a life, to live: infinitive" (237). Stylistically, the definition of Paris tumbles out of the words strung together with commas. Paris means life, and Cummings is also reminding us of the passage on the Zulu, who is a verb, an "is." Any word relating to verbal forms of "to be" refers to C's conception of spirituality and aliveness. Simultaneously, Cummings links abstract thought, language, and concrete place, as in this concluding chapter where he links the verbal to a place, Paris.

Cummings recognizes the web of forces interacting at physical sites and also the mental web that we create out of our experience of place. This "cognitive space," according to David Stea, "is of a different order" than the physical environment and "appears to follow its own set of laws."[30] The brain's functions include not only simple concrete and abstract functions but also "concrete, nonspatially abstract, and spatially abstract."[31] Cummings translates the interactions of physical elements into an even richer cognitive space. He applies this approach with more force and variety in his travel diary *EIMI*. While he used his confinement in La Ferté-Macé to develop transformative ideas of space, his trip to the Soviet Union offers an opportunity to consider imprisonment and freedom in terms of spatial construction and in a larger social context. Instead of considering cognitive and physical space as separate, Cummings looks at the interdependence between cognition and physicality. Rather than assuming that culture imposes structure onto space (as with the Soviet architectural project), Cummings looks at interrelationships, valuing the processes of a living environment.

4

Inverted Space: *EIMI*

"I like it here so much!" lyrically exclaims Virgil "have you noticed a particular feeling in the air—a tension?"

"Have I!"

—E. E. Cummings, *EIMI*

COMMUNIST RUSSIA CHALLENGED CUMMINGS' MANTRA OF ALIVENESS. In the selection above, his tongue-in-cheek sarcasm glides past his naively oblivious guide, Virgil (also called Sibyl and, in real life, Henry Wadsworth Longfellow Dana).[1] Paradoxically, Cummings encountered "tension" in a country hailed for its new freedoms and visited by Western intellectuals interested in the new revolutionary society. His stylized travel diary *EIMI* uses various motifs of containment—shut doors, closed rooms, boxes, cubes, and tombs—and refers to their psychological counterparts—restrictions and delays in his itinerary and uncooperative, suspicious, or even simply tired natives mouthing propaganda. Throughout his month-long trip (May 10 through June 14, 1931), he searched for open spaces and frequently marked his discovery of liberating sites with an exuberant "SPACE!"

In his first novel *The Enormous Room*, Cummings' alter ego C joins with the other prisoners as they transform their living space. In his travel diary *EIMI*, Cummings remains an observer, like Dante, his character's namesake. He refuses to join tourist groups, wanting to choose instead what and who he visits, yet as a hotel clerk quizzically and prophetically remarks, "Do you realize that to go as what you call Yourself will cost a great deal?" (16). As a result of Cummings' chosen self-reliance and the necessity to name himself, he adopts several personas; besides "Dante" he is "Kem-minkz" in reference to the difficulty Russians have in pronouncing his name (21). Nicknamed Comrade K, he claims several labels and travels as "alias peesahtel [writer],alias tovarich hoodozhnik [painter],alias Poietes [poet]" (247).

In *The Enormous Room*, Cummings uses Christian's route as the template for his own spiritual pilgrimage and borrows terminology from *The Pilgrim's Progress*. (C encounters and names modern versions of the Slough of Despair, the Delectable Mountains, and the Celestial City). In *EIMI* Cummings uses Dante's *The Divine Comedy* as his trip's primary organizing reference (although he also refers to Aeneas's trip to the Underworld, calling Russia the "unworld"). Comrade K travels through hell (Soviet Russia) guided by Virgil, a fellow New Englander and a tour guide with serious limitations. He soon meets Beatrice (again, a reference to Dante; she is also referred to as the Turkess or Harem), the wife of the Turk (also called Assyrian).[2] At this point in the story, Virgil becomes an "ex-mentor" while Turk and Beatrice escort Comrade K around Moscow and locate a cheaper place for him to stay than the tourist Hotel Metropole. Cummings leaves Russia via Odessa, where he is befriended by a new mentor, another "Noo Inglunder" (291). He meets his last guide, "dude" (an Afgan), on a boat to Istanbul as he travels out of hell and back into the world.

Unlike the novel, *EIMI* omits specific reference to the route of its prototypical pair, Dante and Virgil, who travel through consecutive levels of hell, purgatory, and then paradise. Cummings enters hell, records his experiences, and then concentrates on getting back to "the World" (377). He finds that while Beatrice resides in hell, she does not belong to it, nor are all the sites which he visits properly hellish (old Russian architecture and landscape in general impresses him as having spiritual qualities). His representation of hell stems from his sense of Soviet Russia as different from the rest of the world. Like Dante, he is shocked at the condition of places and people as well as the poverty and fear that he encounters. His visit to view Lenin's body is as grotesque as Dante's "vision of Satan." Lenin's / Satan's program resulted in major alteration of everything from landscape to architecture to basic human interaction. Nothing works as K expects, such as maps (which makes his anti-pilgrimage difficult). When a Russian fellow traveler tries to show Comrade K their location, the "map-time-table" can only "show us something quite otherwise,i.e.where we all are when" (287).

Cummings's texts rely on similar spatial topoi, regardless of differences in plot structure, and both attempt to deal with issues larger than readers would suspect. Cummings himself notes the significant connection between both works in his introduction to the Modern Library edition of *The Enormous Room*. The goal of his so-called "war" book was to study "something more unimaginably huge than the most prodigious of all universes—[. . .] The individual." He explains that likewise his travel diary is not about communism per se. "Eimi is the individual again; a more complex individual, a more enormous room."[3] Despite his intense dislike of the communist system, Cummings rejects an automatic endorsement of capitalism and emphasizes to his readers that "no artist ever is a capitalist!" (226). "Artist" is the only category in which he allows others to place him, and he consistently refuses to have a purpose imposed on his visit. Norman Friedman argues that "[t]o be for or against war, to be for or against capitalism—these would involve Cummings in those very categories he wants to transcend. The real issues are life against death, the individual against society, and not one institution against another."[4] Cummings establishes himself as an "unhero" by aligning himself against all power structures and with the average individual, including the Russian proletariats with whom he freely shares food and cigarettes.[5]

In both novels, using the central image of the "Enormous Room," Cummings traces relationships between structural metaphors and reality. In Soviet Russia, he wishes that if only a fraction of the people "of a worldless world somehow were not afraid;only did not remember wars;impossibly could forget everything and opening its spirit pour a true structure into space and into time!" (160). Cummings embraces Russians as people who can effect change, suffusing their country with a "true structure." People participate in—and can therefore effectively alter—space through their occupation of space and their participation in the interactions in a field. "Human beings . . . know that they *have* a space and that they *are* in this space . . . for they act and situate themselves in space as active participants."[6] Cummings recognizes that a different kind of revolution will take place when citizens realize the extent of this creative power.

Later in *EIMI*, Comrade K meditates on his ideal of revolution, recalling the Turk's vivid observation: " 'revolution' the Assyrian said 'cannot be ordered like a plate of beans[']" (187). For Cummings, revolution involves individual, spiritual transformation, transcending limitations imposed by our own and others' preconceptions of reality including "predetermined knowledge" and "conquerable space" (186):

> (revolution?
>
> We of whom Is partakes,only to whom our deaths are births [. . .]—breathe in this house,makers of Is,not hope and not despair,but timeless deep unspace,the single poem,which builds unconsciousness:immeasurably this house only shall live! Only these walls may not wilt;only cannot drown with any earthquake,any wave. "Revolutions" everywhere must perish,but not these walls:only these walls are Revolution) (186–87)

Through contradiction, conceptual language, paradox, and mixed metaphor, Cummings approaches a transcendental definition of revolution. Death in and to the world masks a birth to a more spiritual state; as in *The Enormous Room*, the verb "is" represents this transformation to being in motion and growth.

For this passage, Cummings created a form of writing comparable to that of a Zen *koan*. "Koans take as their subjects tangible, down-to-earth objects . . . to make us see, on the one hand, that each object has absolute value and, on the other, to arrest the tendency of the intellect to anchor itself in abstract concepts."[7] Mixed metaphors (the "walls" of revolution will not "wilt" or "drown") isolate and separate the physical properties of the objects mentioned. This "immeasurable" house, where people can breathe (the value of breath is another component of Zen Buddhism), can be described, measured, and conceived in solid form with ceilings, floors, and walls. In the metaphor of the house, Cummings fuses concrete objects and abstract concepts in a startling manner. Zen master Chung-feng observes that the *koan* "cannot be explained in writing; it cannot be measured in reason. . . . For the essentials of complete transcendence . . . nothing can surpass the koan."[8]

Part of Cummings' definition of transcendent experience in *EIMI* concerns the value of space. He repeatedly contrasts conceptions of space as fixed and limited with those of space as fluid and sees the state of Soviet Russia as an exemplar of the repercussions of faulty preconceptions. He remains skeptical of the rationalized claims of communism, viewing the system as incapable of elevating humanity en masse. "People may dare to live,people may be taught or may teach themselves death;noone can learn growing [. . .] Here exists no sign,no path,no distance,and no time. The grower has not any aim,not any illusion or disillusion,no audience" (230–31). When he meets a granddaughter of Tolstoy, who had suffered personal tragedy during the Bolshevik Revolution, Cummings remarks on her continued growth towards inner peace. "Nothing did not matter;everything differently mattered. Values were herself,were not Of herself—not something taken or grasped,but something given" (230).

On this trip Cummings rarely records meeting someone whom he admires and praises. When he does, he finds the quality of their inner being conveyed through outer signs—their eyes, their bodies, the way they move. As with the old aristocratic lady, Cummings similarly admires the spiritual character of a man who buys flowers. "Flowerbuyer" lives in a silent, sunlit room at the top of "shiny stairs"; he moves "daintily, almost sacredly" and speaks easily of the Russian soul (to K's wonderment) (133–34). The quality of his living space reflects his nature. The walk up to his apartment mimics an approach to heaven, and K calls the man "the master" (133). This man loves flowers and the old churches of Kiev. He retains a nostalgic feeling for old Russia that Cummings seems to associate less with political systems than with a highly developed aesthetic sense.

For much of his trip, Cummings assumes that readers understand his value-laden ideal of "aliveness" and instead turns his eye and wit to the failures of the communist revolution to control material reality. When an acquaintance sees a defective sprinkler and remarks "[t]hat's Russia," Cummings ruminates on "how perfectly the far famed revolution of revolutions resembles a running amok streetsprinkler,a normally benevolent mechanism which attains [. . .] transient capacity for clumsily mischievous behavior" (106–7). An example of how the system wastes energy, the sprinkler system also represents the underlying dynamic energy in Russia and its people—the potential for "revolution" to "revolve" and remain in constant, changing motion.

Umberto Boccioni, originator of the aesthetic term "plastic dynamism" and author of several Futurist manifestos during 1910, ascribes Futurism's foundation to concepts of a dynamic universe and the continuous transformation of reality.[9] Cummings reestablishes this Futurist welding of space and action in *EIMI*.[10] His first impression of Moscow involves tactile sensation and awareness of spatial relationships to the exclusion of sights and sounds:

> around,through,under,behind,over myself do amazingly not physical vibrations contract,expand,collide,mesh, and murderfully procreate:each fraction,every particle,of the atmosphere in which moving moves,of my moving,of me,of cityless city,of peopleless people,actually is charged to a literally prodigious degree [. . . .] (21)

To recount the constant motion he feels, Cummings employs directional cues (around, through, under) in combination with verbs of motion (descriptive of precise physical actions such as contracting and expanding). His complete body awareness and sense of motion stands in contrast to "peopleless people" whose identities mitigate their inherent aliveness in who they are and where they live. Yet his experience draws him out of an observer's point of view so that he becomes an integral part of this process "of my moving" and concludes that even "peopless people" participate in the activity. Still, this sense of uniform participation remains an undercurrent of energy, overridden and suppressed by organized revolution on the surface.

Cummings' characterization of Moscow as a "cityless city" points to the inversion of normality in Soviet Russia and relies upon the reader's grasp of the typical qualities of polyvalent space—including the social, physical, psychological, and historical composition of the space in which we live.[11] Cummings develops his own system of identification by using creative descriptions to disclose in minute detail how Soviet Russia has inverted, distorted, and negated the standard structural qualities of architectural space.

Human beings constantly endow space with a host of social, philosophical, functional, and formal qualities. According to Robert Hershberger, our perception of structure includes the "recognition" of form, which we then describe and categorize relative to ourselves.[12] (149–50). People locate many types of representational and referential meanings in architectural structures and infer meaning through recognition of a structure's functions and values. These functions can include human and building uses and physical, social, and psychological purposes. A grand staircase in an opera house, to use Hershberger's example, provides structural support for the ceiling, allows access to the second floor, and, in its wide, gentle incline, serves as an area for social display.[13] Since people evaluate architecture on many levels, a person will recognize a building's use yet also respond "to the forms themselves" as well as to its use or lack of use. Additionally, buildings act as symbols, and "values can be expressed or referred to in architecture."[14] Many of the values inherent in towers, for instance, have endured since medieval times, when a tower provided security and shelter while also signifying power, prestige, and self-sufficiency. Cummings assumes that readers comprehend, through their everyday experiences, the myriad functions of architectural space at stake in his argument.

In "the world" as Cummings calls it, we expect structures to fulfill a variety of standard functions. Doors and windows, for instance, allow air circulation, illumination, access in and out, as well as many psycho-social functions (entering a house often represents a transitional moment; looking out a window allows for private observation of the outside world). In "hell" all these functions are reversed, reduced, and even nullified. As soon as Cummings enters the train from Paris to Russia, he remarks that "SHUT seems to be The Verb" (3). With this observation, Cummings begins the first chapter and introduces a primary contrast between "shutness" and openness ("being"). The train passengers insist that windows and doors stay shut in this "breathless box" (3). Cummings' views out the window of Germany, Poland, and Russia reveal various "SHUTnesses" including "dark fields,smutted towns" (3). "SHUT" represents as much a state of being poor, grimy, and spiritually dead as an action (to shut) and a condition (to be shut). In reference to doors and windows, "shut" describes the effective minimization of multiple functions to a single function, as occupants are restricted to the room and their entrance and exit impeded. Finally, "shut" also indicates a perpetual, inert state of being; Cummings routinely tries to obtain tourist information at Russian Intourist stations but usually finds "INTOURIST shut" to his annoyance (270). Tourist attractions such as the planetarium or Lenin's Mausoleum are frequently "Shut" as well, obstructing his attempt to see Russia (117). The first word of the novel, "SHUT" stands symmetrically opposite "OPENS," the last word of the novel, a word which signifies Cummings' reaction upon returning to Paris and the world (432).

Most sites which Comrade K encounters in the Russian cities of Moscow, Kiev, and Odessa incorporate a denial of their multiple functions, a phenomena which disturbs Cummings and which he indicates to the reader through various stylistic signals. Verbals, adjectives, prefixes, and suffixes serve as identifiers of places and conditions while directing our attention to structural limitations and failures (in the example "shutness," shut + ness indicates the state of being shut; city + less refers to the place minus the values usually associated with it, "cityless"). The root word of an object is often retained as a reminder of that thing's conventional purpose and value, with a descriptive qualifier or negation tacked on. Comrade K stays in rooms which are often "roomless," so that "room" signifies a basic structure while "less" suggests the absence of values and uses normally associated with rooms (117). When Comrade K gets a room in the "hotelless comrade Metropole," he enters by walking up "a flight of marble-or-something" (43). Unlike the rest of the world, basic materials in Russia do not connote structural security, regardless of whether they are actually solid, like marble. In his Kiev hotel room, K sleeps in a "tiny—in complete darkness—bedless" (264) missing all the essential qualities (light, space, furniture) enabling satisfactory occupation.

Cummings finds the phenomena of absent value and suspect materials widespread among the communities of Russian workers. Virgil proudly invites K to visit a "real Russian" living in an authentic "all Russian" home (41). Comrade K satirically refuses to note the romantic quality in living at the end of a "nonlight" hall in a "roomlike room," a "dim hollow cube—3 unbeds,1 untable [. . .] a Shutness,a broken chair,silence,a home made radio" (41–42). Cummings describes structural objects not as physical entities but as either the opposite or absence of thingness. This visit takes K "through Dark to dim,through emptiness to windowed cubic woe" containing nonfunctional beds, table, and a door (42). In other words, he crosses a vacant courtyard and enters a dimly lit hall that leads to a small, poorly furnished, square apartment with a closed door. Cummings encodes these descriptions of structural forms and materials to indicate the level of deprivation in Russian lives.

Men and women receive similar treatment and demonstrate an absence of self-development and bodily substance. When visiting "real" Russians, K notices that the occupants, like their furniture, lack essential qualities, and he mistakes them for "corpses" (42). In general, Cummings uses nicknames for his central characters but lets others fade into the background. Men are faceless "com-rades," women are "nonmen" and "unhe[s]." Instead of proper names, a single characteristic often identifies a person. At one party, Comrade K meets "cadaverous" (83), "Cringecurltwitch," and "pinwheel" (93). This method of naming identifies people as two-dimensional, transient figures.

Thematically and structurally, *EIMI* pivots around contrasts between authentic and false structures. Cummings, after hearing a fourteen-point synopsis of Marxism, "salutes" Karl Marx for creating a complex yet unified program (87). Nevertheless, Marx's achievements remain limited by a mundane conception of human beings and a hasty acceptance of "mere reality" (87). Accordingly, for Cummings, Marx lacks poetry, and his theories succeed in actuating only the material aspect of human life (signified by Cummings as "it is"). Marx's vision omits the more creative, spiritual side of human beings, the ability to dream and imagine (the impulse towards which Cummings indicates with the phrase "I wish").[15] Cummings frets over this willful disregard of humanity's potential:

> with what fatal consequences!for what could less resemble I wish than It is?less an alive actual tree than the "same" merely not dead tree [. . .] less freedom than unfreedom,less dreaming than doing? . . . yet only the bravest fools have glimpsed that difference;with what eyes!
> (now smell earth—no—it's parkless park (88)

Cummings presents his critique of communism as altering nonmaterial aspects of life—freedoms, dreams, and communion with nature—and dramatically reinforces his objections with linguistic cues. He posits that communism is having a pervasive and undesirable effect on all formal and functional levels of life, as the omission of crucial nonmaterial qualities lessens the use and value of space. In essence, Marxism has negated even the basic elements of life, such as the earth, and accordingly a park is transformed into a "parkless."

Red Square serves as the most significant site of conflict between those forces of materialism and those of creativity. They are represented on the square by antithetical configurations of space—Lenin's mausoleum and St. Basil's Cathedral—literally and symbolically vying for spatial control. Cummings first sees the square during Virgil's tour of the city. The square itself offers relief from regimentation and "shutness." As they enter through the square's gate, K and Virgil are "bumped by a most wonderfully refreshing lump of nearly comradeless space" (25). Cummings anthropomorphizes this space as something like a large, furry mutt which takes up an indeterminate amount of room, "bumping" into them. Emptied of people, this space impresses Cummings less as a square than as a nodule, a formless "lump" or mass which itself occupies space. Cummings revels in the experience of humanized space overcoming rigid spatial boundaries.

Cummings objects to architectural forms developed from the traditional emphasis on point, line, and plane (the form of a square, for instance, although in this case the square itself subverts its own rigidity through the quality of its space). In his opposition to traditional design, Cummings essentially objects to the purity insisted upon by High Modernist architecture. Le Corbusier's International Style dominated the International Congresses of Modern Architecture (CIAM) in the 1920s and 1930s. In line with Corbusier, CIAM insisted on "economic efficiency . . . and the universal adoption of industrialized production methods . . . the consumer must readjust to the new conditions of social life flowing from this rationalization" (King, 88). Russian avant-garde architects also led this new order of architecture. After Lenin's death, Moisei Ginzburg argued for "constructivist architecture," with the intention that "buildings, complexes, or even whole cities . . . not only perform their immediate functions but also motivate users (or, if necessary, constrain them) to new actions, new habits—and, thereby, new ways of living" (King, 64). Lenin's Mausoleum represents the erasure of past architectural forms and the reformation of Russian culture.

Cummings' skepticism about this cultural revolution finds expression in its architectural symbols. Standing in the square, Comrade K compares Lenin's Tomb (L's M) unfavorably with St. Basil's Cathedral standing opposite the tomb across the square. He calls L's M a "fake" tomb (244) constructed of inferior, nearly nonexistent materials, "matterlessly inoffensive claptrap,dimly thosethese mausoleum,something,or cardboard or something,or boxes" (59). The tomb is both a trap (as Comrade K experiences during his visit to see Lenin's body inside) and "claptrap", a physical manifestation of the communist program, "the Slogan of Slogans" (25).

When Cummings views Lenin's body (or rather, as he notes, the waxwork replacement), he describes the experience as if he were descending into the pits of hell. He must leave his true self behind and "disguise" himself to enter the building, literally as an American correspondent and figuratively as a man whose "soul descends" (242). Within the building, K becomes entrapped in a series of enclosing forms:

> as we enter The Place,I look up:over(all)us a polished slab reflecting upside(com(moveunmoving)rades)down. Now;a.Pit:here . . . yes—sh!
> under a prismshaped transparency
> lying(tovarich-to-the-waist
> forcelessly shut rightclaw
> leftfin unshut limply (243)

When he is standing in a pit surrounding the body, a slab above reflects the forms of rigid guards and a transparent prism covers Lenin's body. The whole pit is surrounded by a "neckhigh wall" (243). As Kennedy notes, K's description of Lenin leaves no doubt that he is a satanic figure, human to the waist, with claw and fin in place of hands. His shut hand reminds us that he led the country to what Cummings describes as its "shutness."

Architecturally, L's M relies on clean lines and unbroken planes. Inside, basic three-dimensional forms (the prism, the slab, the wall) dominate. On the outside, Comrade K describes the "rigid pyramidal composition of blocks; an impurely mathematical game of edges:not quite cruelly a cubic cerebration" lacking in aesthetic qualities (25). Despite the incorporation of solid forms, the pyramid remains an "aggressively *unreal* boxlike structure" (italics mine, 106). Its excessive material reality results in an unreal architectural composition.

In contrast, St. Basil's Cathedral presents the viewer with an aesthetic challenge to the eye in its curling lines and mixture of colors. The first time K sees the cathedral at the opposite end of the "lump," he stands transfixed; "[t]he tearing of mere me and this miracle from each other demands effort on part of failing benevolence [Virgil]" (26). In previous publications, Cummings had expressed skepticism toward religion; he realizes now that he must change his views.[16] In reference to the slogan "religion is the opium of the people," K responds, "[i]f combining the best(I murmur)elements of barberpole and pineapple be opium,more power to it" (26).

Cummings repeatedly returns to this cathedral, this "sheer barberpolemiracle-pineappleprodigy" (89), in an attempt to explain its overwhelming power. Cummings inserts references to the cathedral as a leitmotif, a reminder of structural principles involving motion, form, and color—all of which are absent from other Moscow sites. He also refers to this cathedral as "the church which makes perpetual revolution"; its visual energy and its architectural form (suggesting movement) stand in contrast to the structural failures of the communist revolution. Overall, the cathedral impresses Comrade K as miraculous:

> Something Fabulous
> a frenzy of writhing hues—clusteringly not possible whirls together grinding into one savage squirtlike ecstasy:a crazed Thinglike dream solemnly shouting out of timespace. (25)

In part Cummings relies on a painterly description of the effect of the cathedral's colors. The eye reacts to "whirls" which Cummings can only describe as one whirl "made of subsidary,differently timed yet perfectly intermeshing,whirlings" (108). The experience of viewing the "Arabian Nights" cathedral incorporates sexual passion, dream states, loss of sanity, and finally transcendence out of the mundane world.

The cathedral 'shouts' out of "timespace," out of the state of when and where (time and place) and into relative *whichness*. K intentionally calls the church a "certainly spiralling Which" in contrast to L's M, which he likens to a hellish place, "a where tremulous with despair and a when luminous with dissolution" (59, 242). Fear, despair, and loss "of omnipotence" dominate timespace; a person who lives outside of timespace lives "freely himself(breathing always round air" and experiences the world in remarkable ways through touch, taste, sound, and smell ("enjoying the sound of the great sun" [242]). Cummings' mysticism leads the reader out of the normal world while always reaching back to normal forms. In the phrase "breathing always round air," he characterizes the air as a structural form—an all encompassing, boundless "round" entity. The cathedral serves as an impetus towards transcendence; it also serves as the focal point of K's escape back to the "world."

As Cummings passes through Kiev and Odessa on his way "out" to Istanbul, space gradually opens up. Countryside, old townships, and finally a view of the ocean mark the progressive stages from shutness to openness. From the window of the train to Kiev, Cummings sees "lush dank vast s-p-a-c-e" (259). He reaches Kiev and a mountain footpath "leads my(finding)self to(suddenly opening immense)SPACE" when he sees a river (264). Through typography and thematic cues, Cummings emphasizes this transitional stage in his travels, as when he exclaims that the cleaner air allows him to "bReAtHe, BREATHE" (264). In Odessa, aspects of the unworld and the world meet, and K becomes disoriented by the contrast of differing experiences. As in Moscow he finds himself in poverty-stricken residences, in one case climbing "through blind stairless unstair nonstair" which appears even more horrific in comparison to the ocean views and wide streets of Odessa. "[Y]ou feel space here," remarks K. The jarring realities in this borderland city cause Cummings some psychological disorientation. (He makes light of his personality split, though. When catching a train, his neighbor jumps on "followed by me and by myself and K and Cummings by all of us rushtearing behind it out through and up" [274]).

Only when Cummings reaches Istanbul does he feel free to piece together his assaulted and fragmented identity. His primary identity as "I" and "C" (Cummings) replaces "Comrade K" and various other nomenclatures. Despite a dangerous nighttime landing and nightmarish wailing from confused passengers, the solid stone dock reaffirms his sense of self in very basic terms ("I. I am. I am standing") and reorients him in the world, "the huge tranquil(w-a-i-l)ity of space" (380–81).

Cummings finds his faith in universal "aliveness" challenged, yet reaffirmed at Saint Sophia. The wailing of passengers dissolves into the wailing of the muezzin. Despite Cummings' thankfulness to be in "the world," he discovers that Mustafa Kemal's westernization of Turkey has affected the people in ways similar to those produced by communism's remolding of Russia. The "old religion" has been relegated to minority status, and Europeanized residents consider Islam's followers to be "dirty and very uncivilized" (399). Cummings once again finds himself aligned with traditional culture and religion.

He has in particular a deeply mystical response to the space, light, and color of Saint Sophia:

Saint Sophia—interior view (from the album *Constantinople, Beyrout, Ba'albec, Damascus,* 1860–1890?).

more skilfully s(ilen)tthan(the
wonde
rfully highest fragrant more)&
beyond joy(be
-yond skill Or and even
si-lence)even(And or total(how
Complex-ly)serene most now of Dreams(i)wan
der [. . . .]
i only tinily now discover universe not)not
height)immensity:and:am;now wonder(joy)&
wander-ing not,floors,but,heavens)

(395)

C interprets the interior as effective beyond the builders' intent and evocative beyond everyday joy and silence; he "wanders" this place as if in a dream. As with St. Basil's Cathedral, this church transcends "timespace" and can only be characterized as "more" and "beyond" reality. The "cUrLiNgS" of Arabic lettering which "squirm" near the top of the dome mirror the whirls on the Moscow cathedral (396). Cummings' eye is also drawn upwards by:

> [. . .] its
> minarets,
> the white shapel
> y against(how-
> blue,SpirEingA
> mong sky(keen:ly;ness the,
>
> (396)

On his first visit, he realizes how the architecture brings things "alive." The value of this space resides in the movement of form and the clarity of color. Cummings emphasizes the fusion of abstract values with concrete substance, as he anthropomorphizes the blue minarets rising "keen:ly" to the sky.

On his second visit to the mosque, Cummings connects his spiritual experiences to poetic reverie. He enters Saint Sophia and finds "inside hugENess,a poet is praying" a "dark poet;blindman:a beautifully communing he" (405–6). This poet's aliveness is "beyond [. . .] dimension;sizeless,entirely creating;impossibly alive:his voice is made of silence and when his voice pauses the silence is made of voice" (406). Cummings renders his impressions of this man and of Saint Sophia in poetic form, which allows him more freedom to experiment with the shape, expression, and pace of his writing.

In linking this architectural space to prayer and poetry, he also emphasizes the nonmaterial qualities of space. The mosque transcends the dimensional categories of space and encompasses both speech and silence in a paradoxical way. Like a *koan*, the experience of being in the mosque lies beyond rational description. As an authentic structure, the mosque conveys aliveness, true expressions of death and love. A central statement on the universal meaning of the structure frames the passage on the minarets. "with its minarets,the white shapely against [. . .] procession eyes,-death;Love" (396).

At the end of *EIMI*, Cummings struggles against the limitations of speech as he tries to explain how Saint Sophia simply prepared him for Paris. His closing reverie builds on themes he reiterated throughout his travels: qualitative silence and speech, fluid space (transcending "when" and "where"), and authentic structures. The "particles" of his self "are not dead are not alive are in time in space" from which he moves ("moveunmoving") "beautifully into everywhere and always" (431). In his recovered paradise:

silence is made of [. . .]
whereless fragrant whenlessly erect
a sudden the!entirely blossoming)
Voice
  (Who:
  Loves;
  Creates,
  Imagines)
  OPENS

(431–32)

This ecstatic ending emphasizes creation and procreation, a transcendent God, and the power of the imagination to create openings. This epiphany involves an emotive form and occurs almost entirely on a spiritual / mental realm. Although his vision of Paris is essentially placeless, he reiterates that he is both within "timespace" and reaching outward. The revolution revolves, finally, to a spiritual plane, but not without transforming human spatial concepts first.

5

Controlling / Creating Space: *Lost in the Funhouse*

"There was all the world I needed; let the real one clip and tumble, burn and bleed. . . . I'd make Mycenaes of which I was the sole inhabitant, and sing to myself from their golden towers the one tale I knew."

"Anonymiad," *Lost in the Funhouse*

I

In much of John Barth's storytelling, worlds open and collapse on one another like unfolding dimensions of a hypercube, disorienting time and space. In "Anonymiad" for instance, the narrator, an unnamed minstrel, sails from Mycenae, geographical site of a historical city (now the site of archaeological digs in Peloponnesos) and center of mythic stories refashioned now by Barth. The story's setting oscillates between real space and myth. The minstrel tells his supposed audience that he leaves his city to see "*The Wide Real World*" and lands by treachery on "*a nameless isle.*"[1] To counter his new isolation, he fictionalizes his life, ironically setting aside the mythic material of minstrelsy to use "actual people and events" in a re-creation of Mycenaes (193). His songs, however, continue to follow archetypal patterns with mundane predictability (we recognize his journey to see the "real" world as a false premise), and they create fictional events to substitute for factual knowledge of events ("to the war and Clytemnestra's treachery I worked out various dénouements" [193]). In the intertwining reality and fiction, how can Barth's minstrel proclaim with certainty that "I'd make Mycenaes"? Can Barth, like the minstrel, make Mycenaes? Who makes a city ultimately?

I posited earlier that when a person enters a spatial field, as when

Thoreau built his house in Walden Woods, that person creates space. They may not necessarily originate the objects of the spatial field or control the interactions within the field, yet that perceiver participates in the field and the meanings it generates. Authorship in this sense is a transitory state. The process of authoring or creating "opens" possibilities (as Cummings would say), and the resultant organization is kinetic in nature. Can we say then that we create space, that we *author* space, when some theorists have adumbrated the death of the author? In considering spatiality, can we even appropriate these terms for a discussion, since they posit as a priori the existence of one creator, one center of perspective? Can we follow Cummings' banner of individualism and create heterotopias for ourselves by transcending traditional spatial boundaries, when Foucault himself decries the validity of transcendental thought? Can we assume that spatial sites can be 'free'—yet interactive and interconnected in what Fredric Jameson (with reference to Lefebvre) calls the postmodern age of fragmentary, discontinuous space?

Henri Lefebvre places us in the age of contradictory space in which simplistic metaphors such as authorship-as-creation and blank-pages-as-material fail and indeed hinder our understanding of the complexity of space. Contradictions such as the ones under consideration here, in which the creation of space occurs without a primogenitor, do not need resolution. We may expose contradictory relationships (and thus fulfill Lefebvre's revolutionary program) while avoiding reductionist resolutions, including the limitation of the subject to an impossible position separate from space:

> 'Human beings' do not stand before, or amidst, social space; they do not relate to the space of society as they might to a picture, a show, or a mirror. They know that they *have* a space and that they *are* in this space. They do not merely enjoy a vision, a contemplation, a spectacle—for they act and situate themselves in space as active participants . . . in a series of enveloping levels each of which implies the others. . . .[2]

Among these enveloping levels Lefebvre identifies various heterotopias: strategic spaces; spaces of consumption, production, and leisure; governmental, global, and local spaces; and spaces of communication (unfortunately, Lefebvre could not discuss the internet, although he certainly anticipates its nebulous space of communication). Finite categorization and systemization of these connections necessarily eludes us in fluid space, so that the definition of a field is somewhat arbitrary. Understanding the process (the production) of space and its interconnectedness precedes the question of authorship, since mapping the process provides clues to relations and power and connections within fields.[3] Kinetic space though, does not necessarily preclude individualism. Lefebvre argues that contradictory space leads to the elimination of forces that attack both individuals and their social organizations. "The body, as a producer of difference—through rhythms, gestures, imagination—has an inherent right to difference."[4] The initiative to claim this inherent right and establish heterotopic spaces accordingly lies with the individual. Cummings certainly accepts the individual's power to transcend given spatial constructs and create spatial relations reflective of that individual power; the question is, does Barth share his views? His works *address* these issues, and at least one critic has argued that Barth *might* have created such spaces by representing the city as both a "site of resistance" and the "scene of replenishment" and "illumination."[5] Yet any attempt to assign an idealistic stance to Barth must confront his postmodern cynicism, factor in his expressions of doubt, and reconcile all this with the increasing preoccupation of his more recent characters with immortality.

Although Barth is more explicit in his explorations of the nature of authorship and fiction than of spatiality, when he does raise the issue of space directly he dramatizes the problematic role of perception and participation. As he tackles the question of authorship in his work as a metaphor for individual creativity, he takes on related questions of perspective, control, and outcome. Unlike Cummings, who hardly apologizes for his celebration of transcendent individuality, Barth—even while celebrating the authoring individual—questions his own assumptions about authorial control and its metaphoric equivalent, individual control. He cynically examines the sources of, and competition for, control and distrusts not only outside forces but the "author" as well. (A frequent trope in his literature is Barth's own loss of control over his text to narrating characters.)

Barth's foremost goal is the subversion of literary narrative traditions, particularly the linear narrative and its companion supposition that each person's life can be understood as a linear narrative.

Likewise, since linear narrative must flow through somewhere, Barth also reconsiders the "setting" of life, the path a person takes in the accomplishment of the metaphorical journey. In short, he critiques the wholesale, historical adoption of these metaphors into Western culture and the subsequent translation of these metaphors into real experience. The idea of life as a journey, as a linear story (having a beginning, middle, and end) with a describable route, has become such a commonplace as to transcend metaphor and become 'things as they are.' By contorting time sequence and displacing narrative control, Barth subverts narrative tradition in books and life. By upending habitual categories of spatial construction—its organization, measurement, and scale—he challenges our understanding of space as fostering a linear narrative of experience. The role of human perception—its center and its duration—lies at the center of this upheaval.

In his first novel *The Floating Opera*, Barth presents a compelling image of metamorphic space as an alternative to the typical metaphors used to describe experiences of three-dimensional space. When "*Adam's Original & Unparalleled Floating Opera*" arrives at dock, writer-narrator Todd Andrews proposes a more original floating opera: "It always seemed a fine idea to me to build a showboat with just one big flat open deck on it, and to keep a play going continuously. The boat wouldn't be moored, but would drift up and down the river on the tide, and the audience would sit along both banks." Usually, since stage and auditorium are stationary, actors and audience share a spatial-temporal realm of experience. In this case, tidal movements and other environmental factors pattern what the audience sees:

> They could catch whatever part of the plot happened to unfold as the boat floated past, and then they'd have to wait until the tide ran back again to catch another snatch of it, if they still happened to be sitting there. To fill in the gaps, they'd have to use their imaginations, or ask more attentive neighbors, or hear the word passed along from upriver or downriver. Most times they wouldn't understand what was going on . . . Lots of times they'd be able to see the actors, but not to hear them.[6]

Actors occupying a movable space of action present a linear narrative that cycles continuously as the boat moves. The audience members occupy various places, and thereby experience radically different plays. The linear sequence of the play, composed ideally of a continuous sequence of parts, becomes a disjointed spectacle. Each individual figuratively steps out of the space of the play, reconnecting with the play for various lengths of time at various places on the continuum of action. In the form of an audience, Barth has created a set of authors who, with their own mental collages, rival the primary author's intent. Yet the inability of these authors to cross the boundary of water and join in or disrupt the play's action relegates them to roles as passive receptacles of sensations. They receive and, to a limited extent, order the story but cannot effectively act in it.

The opera in *The Floating Opera* represents one way to reconceptualize spatial-temporal experience. Barth has continued to explore representations of space-time in subsequent novels, most noticeably in *Lost in the Funhouse, Chimera*, *Somebody the Sailor*, and *Once Upon a Time*. As in *The Floating Opera*, Barth frequently uses fantastic images to critique the reality of spatial relations, an approach patterned after his exemplar, Jorge Luis Borges. Both authors' stories are rife with labyrinths, spirals, bifurcating paths, and geometric figures—all of which are spatial challenges to the twentieth-century ideal of simple perspective, the plane.

Borges leads his reader into mystical conundrums, positing alternatives to traditional space. The city of the Immortals was built by these humans as a "parody" as well as to honor "the irrational gods who govern the world."[7] This city physically exhausts the narrator because of its irregularity, featured in "dead-end corridors, high unattainable windows, portentous doors . . . inverted stairways . . . a chaos" (110–11). A labyrinth is artistic in comparison to this "senseless" city, since a labyrinthine structure serves a specific function, reflected and acted within its structure. (Borges' "mad city" prefigures the work of cyberarchitects, whose computer-generated architecture is dynamic, fluid, and seemingly unbuildable—with notable forays into construction from Frank Gehry and Philip Johnson.) The narrator, trying to establish some kind of motivation for the builders, hypothesizes their imaginative impulses, speculates on their death and the implied antique values of the city, and finally pronounces the city as the result of their "madness." (This critique is echoed by Ambrose in "Lost in the Funhouse," who says that fame, madness, and suicide are the only alternatives to self-deception). The narrator leaves the city only to discover a second kind of immortal space, represented metaphorically by the Hindustani wheel; "on this wheel . . . which has neither beginning nor end, each life is the effect of the preceding and engenders the following" (114). Because of this theory of probabilities, which states that all things can happen in this infinite space-time, the Immortals retreat into Platonic "caves" of "pure speculation" (113). Imaginative invention is posited as the only 'true' reality and "real" life is rejected as incompatible with the evolution of human intellect.

Constructed on philosophical premises similar to that of the city, the Library of Babel encompasses "an indefinite, perhaps infinite, number of hexagonal galleries" which make up the universe (51). In its entirety, "*[t]he Library is a sphere whose exact center is any one of its hexagon, and whose circumference is inaccessible*"; it also comprises "all books" (52, 54). The narrator explains his metaphysical search for the "secret" center, which is "a total book" containing all knowledge, a book from which all other books can be traced (56). He notes that this book possibly exists, excluding only the impossible. In this conundrum the narrator reveals the futility of such a search. "For example [he tells us]: no book can be a ladder, though no doubt there are books which discuss and negate and demonstrate this possibility and others whose structure corresponds to that of a ladder" (57). Books can explicate and mimic, but they always mirror the Other. The difficulty lies in the exclusion of anything from such a book, which cannot function both as the origin of a thing and as the thing itself. Can a stairway exist without a creator? Can such a creator (an origin) exist without needing its own originator? The narrator of "The Library of Babel" suggests that the library is infinite and cyclical—as in curved space, parallel lines will cross in this universe and everything precedes every other thing. God, in the form of a library universe, is perfect and eternal yet "useless" (58).

As does Borges, Barth often considers the question of origins through the topos of authorship. By undermining authorial control, he frees art from the autonomous author. In the seven narrative levels of "Anonymiad," each person's contribution intrudes as another narration; the originator quickly loses control and becomes a part rather than a creator of the action. In "Lost in the Funhouse," when Ambrose gets lost in the back rooms of the Ocean City funhouse, he discovers a man pulling the levers who is a type of Newtonian god, Blake's Nobadaddy. Ambrose determines to obtain this level of control over reality by becoming a "secret operator" of funhouses himself (97). However, Barth's self-deluded narrators do not possess the necessary omniscience to become *fauteurs* of worlds—Ambrose cannot even master the labyrinthine back corridors of the funhouse or the formal intricacies of his own prose.

Despite Ambrose's desire to become a secret operator of funhouses—an autonomous author of life experiences—the narrative veers out of control, exposing Ambrose's amateurish skills as a writer.[8] The story appears to end with the prediction that Ambrose will become a constructor of funhouses, but alternative story endings cast the continuity of the story line in doubt. Earlier, the narrator asserts that "he *lost himself in the reflection*" and also that he "died telling stories to himself in the dark" (94, 95). In subverting the ending of the story, Barth reasserts the primacy of the Möbius strip as a structural metaphor. Seemingly out of authorial control, narrative becomes layered, fused, and twisted, as continuous as a Möbius strip and as compact as a hypercube, and seemingly self-generated.

Both the Möbius strip and the hypercube have been imaginatively "constructed" in architectural terms in order to explore the implications of multiple dimensions in everyday life. These projects "play" with both forms in ways that recall the funhouse experience of Ambrose. In the computer-animated "Möbius House Study," the form of the strip is given architectural form (volume). This "post-Cartesian dwelling is thus neither an interior space nor an exterior form."[9] The resident of such a house might have experiences similar to those described in Robert Heinlein's science fiction story ". . . And He Built a Crooked House. . . ." Heinlein describes a tesseract-shaped house (a tesseract is an unfolded, three-dimensional representation of a hypercube). The house collapses in on itself during an earthquake, trapping the residents in a genuine four-dimensional hypercube. Normal paths through the rooms turn into an Escher-like maze; windows and doors open onto a variety of impossible landscapes. The residents' assumptions of livable space have been recast in seemingly impossible ways. The hypercube erases the meanings of "in" and "out," "up" and "down."[10]

Barth anticipates the disorientation caused by this shift in the cultural paradigm of living space, a shift towards 'plurality' that is amplified in postmodernism in general. Barth's characters struggle

Möbius Strip. Artist's rendering by Eric Olsen.

with anti-linear narration, as does the narrator of "Title," who finds him—or her—self in the middle of a story and struggles with the sense of entrapment which comes from expecting a linear progression to reach an end. Instead, the narrator finds himself in a maze—he can't exit off the Möbius strip—and has lost his sense of spatial and temporal relations. "Beginning: in the middle, past the middle, nearer three-quarters done, waiting for the end. . . . Everything leads to nothing: future tense; past tense; present tense. Perfect" (105). Expectations of traditional time must be set aside in a nebulous present (a suspension of time's progression similar to that in *The Enormous Room* and reminiscent of Gertrude Stein's call for a "continuous present" in *How to Write*). Finding himself stranded, the narrator / author of "Title" decides that "[t]he only way to get out of a mirror-maze is to close your eyes and hold out your hands" (111). In essence, the narrator advocates blind action and the story ends by a sudden act; the last line stops, as if the writer decided to walk away from the problem. "How in the world will it ever" he concludes (113).

Critics often label Barth as an irrealist whose stories "can no longer get hold of any 'reality' at all."[11] Fiction subsumes reality, exposing ontological functions to the reader. "Since we cannot find out what reality is, we shall simply create . . . a world of fiction."[12] Yet Barth does not merely substitute fiction for reality; he uses fiction to discuss the nature of reality, arguing that definitions of reality involve perceptual fields created from experience. The perceiver / narrator of the story becomes entangled in her own plot, and authorship metamorphosizes into the pertinent story. Modern physics poses this dilemma as the Heisenberg Uncertainty Principle, which indicates the uncertainty of an observer's effect on an experiment, given that even observation becomes participation. As Barth explains in a lecture series (published in *The Friday Book*) entitled "4 1/2 Lectures: The Stuttgart Seminars on Postmodernism, Chaos Theory, and the Romantic Arabesque," his interest in spatial-temporal themes goes back not simply to Borges but to his readings of new theories in physics.

In response to critical viewpoints that still espouse the need for a fundamental basis of reality, Barth offers the möbius strip metaphor. At the beginning of *Lost in the Funhouse* is a "Frame-Tale," an unfinished Möbius strip printed with instructions to cut along the dotted lines, twist, and glue together. Steven Bell, among other critics, notes that the arrangement of stories in *Lost in the Funhouse* resembles a Möbius strip: "[L]ike the Möbius strip, [the book] is not just a circle, but rather a circle with a twist; the book does not circle back upon and close itself so much as it is open-ended, or rather open at both ends. From this point of view, if the book could be said to 'close' at all, it would be only (paradoxically) at its center, an artificially static moment . . . in an essentially dynamic process."[13] The structure of the book and the expectations it imparts place value on process (the dynamic of unending twists) rather than on artificial designations of beginning, middle, and end.

The strip represents in part the conflict between the finite and the infinite, yet it is still a piece of inked paper existing in three-dimensional reality. Critic E. P. Walkiewicz interprets the motif of the Möbius strip as asserting the foundation of three-dimensional reality. To Walkiewicz the strip reveals basic structural form in its material construction. "Discovering that the strip is not multi- but merely three-dimensional, we may also be led to discover the inviolate level that exists below every hierarchy."[14] Thus Walkiewicz transforms Barth's critique of the search for origins (a creator or author) into the search itself. He does not consider the alternative, that when we envision beings living on the strip, every stroll around the strip, as Michio Kaku explains, results in an inversion of our bodily components.[15] This latter assumption places us in the realm of science fiction but also reinforces the need to examine the impact of movable points of perception. The potential for multiple centers of perception and participation (as Barth also shows with the spiral in *Chimera*) challenges reductionist philosophy and the idea that an "inviolate level" necessarily exists.

II

In "Lost in the Funhouse," Ambrose's day at the Ocean City Boardwalk turns into an investigation of the basic assumption of structural integrity at this heterotopic site. He literally looks beneath the boardwalk and behind the walls of the amusement centers to see the mechanisms which power the flashy social exterior. Because Ambrose changes his point of perspective, he can explore the workings of this typical summer resort. He takes the initiative to question and think about the site's functions and purpose, thus exposing the "illusions" of this site which is already busily countering the illusions of everyday, working life. He questions the purpose of the family vacation as well, the relatively new phenomena of "leisure" time in industrial society. As the narrator / author, Ambrose also second-guesses his own role as a perceptual center of this spatial field, the boardwalk.

Unlike the creation of an imaginary Mycenae, Barth not only begins with a real site, he stays there and proceeds to critique its physical structures and social functions. He keeps the juxtaposition of physicality with social dynamics at the forefront of the story. In understanding Ambrose's immediate critical reaction to the boardwalk, the reader first recognizes the given complexity of this site. How does the boardwalk fulfill its cultural function as a "countersite" to everyday spaces? Boardwalks provide areas for unrestricted fun in contrast to daily life and allow visitors freer personal expression than society normally allows. The idea of the boardwalk metamorphosed in the late nineteenth-century out of the festival, an outdoor event that led to more specialized sites of recreation and entertainment such as the amusement park, the theme park, the neighborhood carnival and street fair, and the county fair. These festival sites allow visitors to play games, enjoy entertainments, and mingle with less social restrictions. Amusement parks originated from the advertised goal of providing relief from the puritanical customs of Victorian society and the mundane work routine. Edward Tilyou, son of Coney Island owner George Tilyou, remarked that amusement parks let people "cut loose from repression and restrictions, and act pretty much as they feel like acting—since everyone else is doing the same thing."[16] Parks, fairs, and other festival sites provide the illusion of social freedom in a temporary setting.

The boardwalk fulfills Foucault's criteria for a heterotopia: it juxtaposes several "contradictory sites" (an idea which mirrors Lefevbre's concept of contradictory space); it provides a "break" with "traditional time" (26); it includes a system for entering and exiting (thus marking the boundaries of the special place); and it functions as a "space of illusion" and "of compensation."[17] Most importantly, it is a safe environment for experimentation. Despite the transformation of the calm and stable neighborhood into a busy, transitory place of vacation, the boardwalk still operates by consistent rules of organization and behavior. It offers a safe place that challenges but does not threaten the everyday social order. Ambrose's (somewhat nervous) observation holds true for the entire boardwalk as well as the funhouse: "it's meant to be a *fun*-house; that is, a place of amusement. If people really got lost or injured or too badly frightened in it, the owner'd go out of business. There'd even be lawsuits" (90). Heterotopias operate within the bounds of social expectations.

An actual place still in operation, the Ocean City Boardwalk, a complex site combining architectural structures, natural spaces, and man-made paths, was begun in the 1880s and provides several types of amusement still popular with vacationers: "The amusements [were and still are] located on the southern end of the boardwalk . . . bowling alleys, shooting galleries, Japanese bowling, merry-go-rounds, fortune-tellers, funhouses, and games of chance . . . concessions encouraged a carnival atmosphere. Today the boardwalk extends approximately three miles to the north of the inlet."[18] Daniel Trimper's "Windsor Resort," also named "Luna Park" (possibly referring to Coney Island's Luna Park), occupied part of the southern end of the boardwalk.[19] Trimper's famous rides included the oldest continuously operated merry-go-round and a ferris wheel. Another entrepreneur, Mr. Jester, built a funhouse in the late 1920s and placed the mechanical dummy "Laughing Sal" (Barth's "Fat May") in front of it.

Permanent residents maintain the hotels and shops of this resort town while vacationers arrive "to shed routine activity, dress, and sometimes good behavior."[20] As John Kasson notes about Coney Island in New York, such places "promised not repose but activity, not stability but transformation . . . visitors exchanged the dingy banality of city streets and the pompous sobriety of public buildings for a colorful holiday world."[21] The extravagance of amusement parks and boardwalks suggests certain aspects of contradictory space in which color, facade, and decoration oppose the more restrained resources of the working middle class.

Amusement parks mimic the streets and buildings of an urban neighborhood, and the ferris wheel and roller coaster provide a horizon of tall structures like a city landscape. Both boardwalk and park reverse the norms of family life in a single-family home. The boardwalk mimics a variety of community conventions. Sidewalks no longer serve as paths for pedestrians to get from one place to another; rather, the walk provides an aimless space for socializing, eating, and people-watching. Ocean City provides a place for family outings, transforming the single-family homesite typical of most neighborhoods into funhouses, arcades, and hotels offering temporary housing. In contrast to the family meal, concession stands offer sugary, fat-laden, low-nutrient food. The family hierarchy, headed by the father, is encouraged to dissolve into childishness as everyone plays games. "In Coney's permissive environment customers felt a giddy sense of irresponsibility."[22] On the boardwalk, "the shrieks of gulls merge with the clatter of thrill rides and the barkers in game booths . . . [it seems] this is a town of perpetual youth."[23]

The Ocean City Boardwalk juxtaposes several contradictory sites that accommodate a range of activities, including the beach, the hotels, the places of amusement, and the boardwalk and pier itself. As a festival place it changes the nature of time by suspending the normal accumulation of events in history; the working world pauses for a period of relaxation. Ocean City is also set apart from everyday life by a physical barrier—vacationers must first cross the inlet to reach the barrier island—that signals the point at which time and place change meaning. The boardwalk and individual amusement centers, by posting fees and setting up gates and fences designating entrances and exits, mark the site as special. People must also carry money to gain access to various places on the boardwalk. Finally, as mentioned previously, it is a theatrical and gregarious place. "It is a [place] of illusions—illusions of fun, illusions of progress, and most of all, illusions of people about themselves and how they look in shorts or bathing suits."[24]

The complexity of Barth's story lies in the allusive interwoven references to literary representation, material reality, and personal fantasy. Barth mentions John Dos Passos having written about the early train travel to Ocean City; this reference leads Barth's characters to recall "the steaming trains and stuffy long dresses" (73). Barth also uses historical details to draw in references to the real Ocean City, as when he describes the adults standing "at the end of the boardwalk where the Hurricane of '33 had cut an inlet from the ocean to Assawoman Bay" (86). His characters, also recalling various historical and cultural details, take these layers of references and formulate subjective viewpoints. For Ambrose's father the replacement of train travel by automobiles symbolizes "part of the deterioration of American life" (73). By demonstrating how people overlay their own opinions and perceptions on historical reality, Barth questions how we can know reality in any pure form.

The prose style of "Lost in the Funhouse," with its authorial intrusions explaining literary and grammatical conventions highlights the artificial nature of those conventions of realism and suggests that conventions substitute form for the thing itself. When Barth uses an eighteenth-century convention, printing blanks after an initial to indicate real street names (such as "B____ Street"), he draws our attention to the artificiality of this once standard device of realism. After mentioning the hurricane of 1933, he returns to the question of authorial control over realism, double guessing his use of facts; "was it Assawoman Bay or Sinepuxent? Are there other errors of fact in this fiction?" (88). Despite his re-creation in this story of the concrete details of place at a certain historical time period, his details enter the story at awkward moments in the narrative, unconnected and sometimes expressed with uncertainty. In describing what lies under the boardwalk, Barth begins with a detail and ends with a vague suggestion: "matchbook covers, grainy other things" (79). His own experiences at the Ocean City Boardwalk as a boy—memories which he obviously draws on for this story—conflict with his notion that translating reality to fiction poses certain problems. He describes the boardwalk in detail, yet his awareness that he is manipulating reality provides a parallel to Ambrose's own difficulty understanding and participating in his family vacation. Ambrose finds himself in a real place struggling with hallucinations, fantasies, and personal, alternative readings of the boardwalk site.

Barth's concrete details reinforce our sense of material reality in juxtaposition to the uncertain narrative progression of the story. We read about the fortune-teller machine, whose gypsy was "cracked and taped, her kerchiefs and silks long-faded" (84–85). Sailors on leave play games and visit the "penny arcades and shooting galleries," while couples walk past the funhouse and Lover's Tunnel, talking and watching others (86). Boardwalk strollers can enter an area "set aside for ferris wheels, carousels, and other carnival rides" including "the vast and ancient merry-go-round" (88). As a heterotopic site, the boardwalk site contrasts with daily social routine by creating "a space of illusion" that "exposes" the illusions of other places.[25]

Barth's interpretation of Ocean City restructures spaces already reconfiguring normal social customs, thereby critiquing our standard assumptions of heterotopic sites. Ambrose resists the given purpose of the family vacation and interprets the boardwalk and its amusement parks in innovative ways. He suggests that "[i]f you had x-ray eyes and could see everything going on at that instant under the boardwalk and in all the hotel rooms and cars in alleyways, you'd realize that all the *normally* showed, like restaurants and dance halls and clothing and test-your-strength machines, was merely preparation and intermission. Fat May screamed" (89). Ambrose sees the boardwalk as a place of empty spectacle, which, instead of providing vacationers with a qualitative real space, hides more important spaces. Rather than joining the crowd passing under a laughing Fat May, Ambrose hears Fat May scream and feels threatened by the boardwalk's misrepresentation of space. Later, he superimposes a winter vision over the summertime boardwalk scene. "Ocean City's disserted, the ghost-crabs are tickling across the beach and down the littered cold streets. And the empty halls of clapboard hotels and abandoned funhouses . . ." (86–87).

In trying to uncover the "real" site, Ambrose inadvertently deceives himself, in accordance with the spatial distortion he senses and replicating also the larger social deception in which people participate on the boardwalk. The crab becomes a symbol of this delusional avoidance of deceptive appearances. An old-timer advises Ambrose to move "crab-wise" through the funhouse barrel in order to watch others fall without falling himself (92). Yet with this advice, the old man becomes a "betrayer," one who misrepresents the site's purpose. Ambrose, moving into the mirror-maze, recognizes himself as a betrayer as well through his own self-deception (93). Then the funhouse itself betrays him when he cannot find the exit door or the funhouse operator.

Although the boardwalk and its amusement parks contradict the regular functions of everyday sites, this site still operates by its own set of conventions. An amusement park is an "engineered environment, carefully planned to manipulate visitors into having fun but also spending money."[26] It serves as a family "play field . . . a place of action, noise, color and confusion which people enter only to play."[27] Ambrose misinterprets this function and becomes an "intrusive" voice during the family vacation, much as Barth intrudes into Ambrose's narrative. He "took a wrong turn" and finds himself socially "outside" the site, while still physically there (95). He gets lost in the heterotopia as he does in the funhouse, finding himself still at the site but on the wrong side of the walls. By stepping outside the conventions of even an unconventional place, the funhouse stops being fun and becomes "*a place of fear and confusion*" (72).

Ambrose's memories also betray him, causing him to question the connection between what he thinks he experiences and objective reality. Although he originally describes the operator's room in realistic detail, with its "crude panel of toggle- and knife-switches hung beside the open fuse box," he later wonders if "he hadn't dreamed part or all" of the back room (87). The changeable nature of the funhouse, as he experiences it, threatens his sense of self. Ambrose loses his identity just as he drops his "name-coin" in the revolving barrel (92).

In the story, Barth creates a "chain of levels of reality" with multiple authors: Ambrose describes a funhouse (fictitious Ocean City) in a funhouse constructed by Barth (the book) copied from a funhouse made by builders (the "real" funhouse) in a funhouse (the world) constructed by God.[28] If, like Ambrose, we step "outside" the world, will we also face a mirror-maze, a loss of identity like Ambrose's dropped name-coin? At first Ambrose recognizes only three "alternatives" to self-deception: "fame, madness, suicide" (93). Then he discovers a fourth alternative and hopes that he can become a creator, an "operator" of funhouses, someone who steps out of bounds yet still participates in the game as an organizer.

With this conclusion, Ambrose enters another phase of self-delusion. As the narrator of "Life-Story" asks, "what sort of story is it whose drama lies always in the next frame out?" (121). To end the vertigo of endless "et cetera et cetera et cetera" which the "Life Story" narrator also experiences, Ambrose arbitrarily chooses an end frame, the operator. However, embracing this sense of infinitude actually "ends" the story, as it did in Borges' "Library of Babel." In "Life Story" the narrator posits that either he must continue to wrestle with "the old analogy between Author and God" or "establish some other, acceptable relation between itself, its author, its reader" (128). *Lost in the Funhouse* poses this dilemma repeatedly and asserts the necessity of finding some "other" answer to the problem of origination.

Heterotopic sites often compensate for the authoritative relations between place and occupant in everyday space, since they function to expose illusory social-spatial constructions. Architectural theorist Ross King interprets Foucault's intention in defining heterotopias as meaning to "expose the inadequacy of our lives and their spatial settings . . . *to deconstruct the social production of heterotopias.*"[29] In a sense, Ambrose deconstructs his family vacation to Ocean City, serving as the funhouse mirror to this heterotopic site. He looks under, around, and through the spatial configurations that everyone else accepts and joins unquestioningly. Like Cummings in Soviet Russia, Ambrose asks these questions for other people. His youth and inexperience threaten to trap him in solipsisms, yet he does not shirk from "alternative" modes of participation. King calls for new motivating visions of what life can be—"heterotopias of resistance and remotivation."[30] The act of transcending the social conventions of established heterotopic sites serves as one facet of this resistance. Asking questions about the value and function of architectural sites unfetters the impulse to recreate these sites.

III

Ambrose becomes enamored with "the old analogy between Author and God" as a paradigm for 'authoring' his own life; however, his experiences at the funhouse suggest another solution to a deceptive appearance of control. Although he posits only "fame, madness, suicide" and "operator" as the four modes people can live in, his actions suggest a fifth—exposure of the 'inadequacies' of social spaces and reexamination of the self's role in the creation of social spaces. Barth continues this examination of assumptions about spatial organization ("reality") in successive books. In considering the need to understand the creative process without an "operator," Barth develops the spiral in *Chimera* as a companion metaphor to the Möbius strip. The "Perseid" section of *Chimera* describes a completely imaginary temple in the shape of a spiral. Barth uses this imaginary space to critique the relationship between real and imaginary structures and question the validity of the self's adoption of a centrist spatial perspective. Perseus finds himself lying on a couch in the center of a temple with the nymph Calyxa, who affirms that they are in heaven.[31] Unlike the Greek temples that Bruno Zevi describes, with closed interiors and religious activity taking place outside and around the temple, the action of "Perseid" takes place almost entirely within the temple. Contrary to the supposedly ethereal nature of the place (Perseus later finds out that he is alive, not dead) and its obvious nature as a fictive invention, Barth provides an extreme amount of realistic information on the temple's structural form. Perseus describes the temple from his couch:

> a velvet gold rectangle with murex-purple cushions . . . unwound from my left-foot corner in a grand spiral . . . upon its walls curved graven scenes in low relief, each half again and more its predecessor's breadth, to the number of seven where the chamber wound from view—which scenes . . . I saw depicted alabasterly the several chapters of my youth. . . . (69)

The spiral of the walls mirrors the spiral descending within Calyxa's navel; her navel, "rather than bilobular or quadrantic like the two others I best knew, was itself spiriferate, replicating the infinite inward wind both above and below the finite flesh" (76). The temple continues to twist outward, its walls enlarging proportionally in height and length as the temple's spiral unwinds. Because the temple walls depict events in Perseus's life, the temple structure incorporates the passage of time, made visible in the artwork and in the structure itself. As the temple walls lead outward, the panels "speed up" and the period of time between pictured events shortens.

The precision with which Barth describes this temple belies the emptiness assumed to exist in an ancient temple that contains space reserved for the occupation of an omnipresent god. Rather, this temple contains real people between stone walls depicting Perseus's life story, originally recorded by his biographer and illustrator Calyxa. She assigns each panel a precise coordinate for reference; she tells Perseus, for example, that he will "meet Medusa in Series Two, Mural F, Panel One" (103). At this point, issues of realism, artistic depiction, and mythic dimensions become conflated (as often happens in Barth's fiction). Barth asserts the temple's structural realism, despite its connections to myth. He envelops his entire story in imaginary talk among mythic constellations (Perseus and Medusa, with Calyxa represented by a spiral nebula) in a paradoxical comment on realism and myth as interacting systems of equal value.

Perseus worries about the seemingly endless succession of panels and the decreasing amount of time between represented events. "Mightn't it be, then, that like the *inward* turns of the spiral, my history would forever approach a present point but never reach it?" (111). Like the "Title" narrator / author, he fears that his story will never reach an end. Although the temple's structural form represents the double infinity of a spiral which twists inward and expands outward, a situation which makes Perseus wonder if he'll ever view his own death, Barth places his structure in the desert and reveals that this infinite spiral actually ends at an entrance. Calyxa finally admits that Perseus is "not in heaven" and that the last panel doesn't exist; it's "just the desert, as you'll see on your way out." (117–18). The temple represents real space and time, and instead of remaining trapped in infinity, Perseus is released into the real world. He simply gets up and walks out.

For Barth, the figure of the spiral indicates constant movement, both inward and outward. As a spiral, the temple allows the resident to perceive space in multiple directions. Barth reemphasizes the dynamics of the spiral through the temple murals—artwork that takes up discrete yet successive blocks of space in order to represent units of time. As the original inspiration of this artwork, Perseus can elaborate on each panel's picture, thus expanding the depth of the representation through the activity of his memory. As with Adam's Floating Opera, the artistic description of events relies both on the structural form of the artwork (whether murals, drama, or prose fiction) and on the conditions governing the audience's perception. Although Perseus re-narrates his life for Calyxa (and narrates her artwork for us, the readers), he leaves the narrative and its insular temple world and becomes transformed into a constellation when he looks at Medusa. Barth reminds us that mythic stories originate from the world we live in and the actions we take but also that structured, narrative control is illusory and bound to be cycled into some other pattern.

When we choose, however arbitrarily, our subjective "narrative" center, as does Perseus, connections to other fields spin outward in multiple directions. Our sense of center changes as we move through space and interact with other centers of activity. Although

Foucault does not emphasize the dynamic movements of the perceiver, movement is a key component in the organization of a spatial field. Theories of dynamic sites have replaced the Renaissance ideal of orientation around a fixed, subjective center, which provided the subject with "an immobile perceptual field, a stable visual world."[32] Heterotopias counteract the debilitating effects of perceptual stagnation, and society has created these places to challenge and unfix our habitual perceptions and activities. In a heterotopic space, other sites "are simultaneously represented, contested, and inverted"; they "reflect and speak about" the everyday world.[33] While Barth's temple world 'reflects' the everyday world of Perseus in the narrative, it only 'contests and inverts' that world through its spiral structure, which eventually lands Perseus outside of its walls and its mythic space of "heaven" and into a counter-site, paradoxically an equally mythic 'real world.'

The concept of dynamism changes the dialogue about self and world in Barth's next book as well. While Ambrose's perspective challenged the illusions of the funhouse (revealing its static nature and entrapping features), the spiral temple demonstrates the necessity of dynamic space as offering a way past the limitations of singular perspective. In this context *The Last Voyage of Somebody the Sailor* reconsiders architecturally structured space as a vehicle for transcendence, examining in particular the motif of the ship and the idea of voyaging (moving through space) as an example of dynamic space and as a metaphor for transcendence. *The Last Voyage*—often identified as the mirror text of *Chimera* in its use of Arabic mythology as the latter uses Greek mythology—is more directly "about" the individual's sense of center, interconnectedness, experiences of transcendence and, in Barth's terminology, "translation." Unlike his other works, this one deals more explicitly with issues of transcendence (although the specifics of this dialogue are still relegated to a few short passages). To complement his discussion of "crossing over," Barth deals thoroughly, as he does in other works, with issues of realism and perspective, in this case by juxtaposing realistic Maryland (rendered in equally realistic prose) with the fantasy world of Sindbad the Sailor's Baghdad. Several heterotopias, set in these two locations, are represented and either discarded or transformed during Somebody's search for transcendence. Echoing many themes and questions from *Lost in the Funhouse*, *The Last Voyage* never commits to an "answer." (We discover that the narrator of *The Last Voyage* may be either insane or dying . . . either option destabilizes the message of that narrator, and, as an unidentified "somebody," he certainly isn't famous). Yet the novel suggests a framework from which to conceptualize space and self, reality and creation.

As with Barth's previous texts, indeterminate sources of authorship unseat the narration of "reality" and "facts," highlighting the effects of perspective as an unfixed, unreliable center conjoined with other centers of perspective. The narration runs at least eight narrative levels deep (depending on how you count them), turning on itself like a snake swallowing its tail. Barth writes from the point of view of " 'Baylor' the Taler of Behler the Failer" who tells Scheherazade's story (to Death, her "familiar stranger") of Somebody's last voyage. Within this narration, Somebody the Sailor (also referred to as Sinbad the Landsman, the Still-Stranded, and the Shipwrecked) relates stories of his life as Simon William Behler. Somebody tells these tales of his voyages in public to the guests of Sindbad the Sailor and in private to both Sindbad the Sailor and to his daughter Yasmīn. Within this narration, Simon B (for Behler, also known as S. W. Behler, the journalist) mentally narrates his life to his twin sister, deceased at birth. He alternately narrates (as "William Baylor" the author) to an undetermined readership via his travel log. Lodged in between Somebody's narration are Sindbad's narrative of his voyages (both public and private retellings), Yasmīn's retelling of Simon's stories through sexual games, Kuzia's "gloss" on the stories rendered in the physical gestures of a belly dancer, and Jaydā's narrations of her own "story" which she claims to be the central part of "Allah's great story." Not to mention that within the structure of this entire narration reverberates the "official" voyages of Sindbad the Sailor of *The Thousand and One Nights*. As readers, we experience this continuum of subjective viewpoints involved in retelling and note the improbability of the narrator's assertion that Scheherazade knows the "official version."[34]

For Somebody, the narrative purpose is to solve the puzzle of how he was physically moved from one state, time, and place (the "real" contemporary Indian Ocean where he was sailing on his and Julia Moore's *Zahir*) to another state, time, and place (the "mythic" ancient Persian Gulf where the crew of Sindbad's *Zahir* rescues him). This immediate puzzle becomes entwined with

Behler / Baylor's progressive loss of his sense of identity and his ongoing search for his "real and first and final name" (573). As in other Barth stories, this narrator needs grounding in all senses of the word and some idea of truth. Within alternating chapters of the novel (the "voyages" and the "interludes") Maryland and Baghdad 'reflect' and comment on each other, and this pivoting between places also decenters the narrative. The principle narrators, Somebody and Sindbad, are paired as well, with Sindbad serving as a "straight man" to his namesake, the time-and-space-traveling Somebody. Sindbad remarks at the outset of the tales that the two men will "rehearse their separate voyages one by one . . . till it be discovered how they interweaved and ended at their common starting place: right here, right now" (24). Sindbad's naïve assumption that this complex series of events, places, and times can have a "starting" point is undercut by his companion's narrative.

Sindbad tells stories in a satisfying way, as an after-dinner apéritif, featuring action scenes and easily recited lessons. Somebody's stories, although dissatisfying to Sindbad's guests (who are used to narrative fantasy), probably satisfy most contemporary readers, who are used to realistic prose narrative. When these stories are placed together, however, the resultant comparison challenges us to rethink definitions of narrative story, linear plot, the metaphor of life as a narrative or voyage, and the 'reality' of time and space. In "Somebody's Fourth Voyage," Julia and Baylor dismiss the essentially fake historical project of Tim Severin's Sindbad voyage. Through exacting replication, Severin's modern-day crew hopes to avoid reminders of the "Here and Now." For Julia and Baylor, the "Here and Now is inside our Heads as much as out there" (326). Although at times Barth seems to be arguing for a centric, Kantian role of the 'inside'—the organizing perspective of the self—he counters with the assertion that 'inside' and outside are interreliant entities. As Sindbad remarks, "a dream long dreamed may grow a real geography" (26), transcending the boundaries between solid, physical reality and imagination.

To the fourteen-year-old Baylor, the idea of transcendence means that the "I" becomes "the very lens of the cosmos," at which he experiences the "ultimate transfigurement" (107). The focus on "I" and the equation of "I" and the universe places him squarely within traditional concepts of transcendentalism. Yet, he's not sure exactly what would happen next when he reached this disembodied state . . . "Levitate? Evaporate? Melt? Cross over?" (108). His articulation of transcendent experience becomes more defined as he ages and his definition of transcendence broadens. With a new sexual partner (and possibly in a drug-induced state), he experiences "the elasticity of time and space . . . *that swooning sense of being not "out-of-body" but supernally* in *it, beyond the particulars of time and place*" (200–201). In a near-death experience during a cruise in the Virgin Islands he finds himself "suspended between worlds" and unable to completely experience either place, until he breaks through the "interface" (233). Next in this series of experiences, Baylor 'looses' himself by taking "a wrong turn" in Charlotte Amalie. His interpretation of this sudden visit to "Marakesh" is that he did not experience "an 'altered perception' but a quite ordinary perception of an unaccountably altered reality" (307). The physical reality of this translation—an experience akin to Thoreau's insistence that he feels rocks and mud underneath—foreshadows Baylor's physical transport to mythic Baghdad.

Once there, Somebody and Sindbad compare voyages, and this comparison points to their different approaches to the search for an ideal world (an idea given amorphous shape in the image of the "magic isle" Serendib). Sindbad sets out on each voyage knowing that he will lose his ship and that, if he brings his "tub of last resort" and acts smartly on lessons previously learned, he will survive and return home richer. The value of "home" is reinforced by leaving and returning (behavior that mimics one of the ways in which a site is "heterotopic" in having a special entrance and exit). For Somebody, the return "home" to Maryland, or Serendib as the Arabs call it, occurs "serendipitously." As he explains to Sindbad, referring to a trip which involves temporal-spatial transcendence, one has to "sail sidewise" in order to reach Serendib (13).

Why does one man succeed in finding Serendib and the other fail? Somebody, who enjoys the "Here and Now" yet privately longs for transcendent experiences (or sometimes, just really good sex), "serendipitously" relocates himself in time and space on three occasions. Sindbad fails to create a haven for himself despite multiple attempts. He fails to transcend everyday life despite his restructuring of his surroundings, the goal of which was to rework space and achieve a setting more in line with his adventurous life. To a limited extent he succeeds, but his successes mask other failures. Sindbad's grounds are transformed into a desert in order to deflect his increasing hydrophobia, yet his garden only manages to simulate desert isolation and dryness (despite the realistic touch of a roving camel). His home too undergoes a complete restructuring, in that he substitutes tubs for the various functions of his living space. One structure replaces another, so that the refurbished site of the tub does indeed 'reflect and speak about' both its former function and the room it replaces. The tub is the most complex architectural structure in the book, even more so than the other central structure, the ship. As a metonymic and metaphoric sign, it is rivaled in complexity only by "Serendib," a word simultaneously signifying mythic, geographic, and lexical meanings.

The tub is a unique heterotopic structure (Barth's addition to the original Sindbad stories) whose illusory meanings are progressively exposed. The tub's original function for laundering has been replaced, and its structure modified to accommodate new functions. Sindbad's various tubs serve as miniature housing, providing a place to eat, sleep, sit, have sex ("tubbings" is a clever pun on "tumblings"), talk, dance, and even striptease. The tub's first, new function was to serve as an alternate form of travel (to replace whatever ship is sinking underneath him). Like other architectural structures, these tubs convey a set of values, referenced in their names as the "Tub of Truth" and "tub of last resort." In the same way that a building accumulates physical, social, and psychological meanings, the tub takes on allegorical and metaphorical meanings during the narrative. The 'bedroom' tub is vaunted as a site of "truth," reported as the site of oracular 'readings' by Jaydā and valued by Sindbad as a protective totem. Yet the "truth" falters under contradictory reports of Sindbad's fifth voyage, the interlude during which Kuzia complains about the "Tub more of Artful Simulation than of Truth" (382). Even Sindbad, having lost the 'original' tub, disenchantedly refers to its replacement as his "Tub more of Verisimilitude than of Truth, and of Next Resort more than Last" (388).

From beginning to end the tub remains a site of contradictory meanings. Inverted, it serves as a throne from which the Caliph dispenses judgments, so that the dinner guests get their wish that the narratives and their respected narrators would be "steep[ed] in the Tub of Truth . . . to scrub away false cosmetics and true discrepancies until their veritable complexions are revealed" (433). Not all truth will be revealed in this narrative, however, leaving readers and audience to guess at the incestuous relationship (never named as such) between Yasmīn and her father. The tub *appears* to be an "open" structure, literally and metaphorically. Its widest "side" (the top) reveals its contents, yet its occupant, Sindbad, subverts the structural intention of easy access and view. Even Yasmīn's leave-taking from her father, as she climbs beside him within his two tubs, balanced on his camel, is well hidden from view. "You see nothing" protests Yasmīn. "You see a dutiful ex-daughter bidding her ex-father farewell forever" (361). A contradictory site of truth and falsehood, wish fulfillment and banishment, conflicting family relations—the tub represents the seat and unseating of power.

Despite the tub's role as a site of secrecy, lies, and double-dealing, its open structure allows Somebody to climb in and out at will, disrupting the integrity of Sindbad's power base. In comparison, this free activity is closer to Somebody's conception of a site of transcendence, which is ultimately structureless. For him, such a site is a "way station" (440) and transport through it means crossing the "boundary," leaving structural security and altering place and time. What Somebody seems to discover through these events and his subsequent return to Maryland is to value the experience of boundary crossing. His ship ultimately fails to carry its human cargo and sinks, yet Somebody continues the voyage. He transcends the container, continuing the action for which the container was built and arriving in a new place alone and naked. The elemental freedom of the body in its natural, physical surroundings serves to comment on the function of human-built structures—and while these structures fail to provide a permanent space for transcendental experiences, they serve as conceptual models for human activity. The ship represents the metaphoric voyage of life, the element of *process*. Eventually, though, the structure itself becomes confining. Only when Baylor / Somebody loses everything (partner, ship, clothes), does he manage to "cross over."

In a sense, Baylor exchanges one idea of transcendence, which focuses on the "I" as an *a priori* agent, for another, twentieth-century strain of transcendental thought which borrows the concept of loss of self and world in part from Zen Buddhism. He discards the idea of self as a static container and instead participates in transfiguration. As a representative of this event the *Zahir* is serendipitously named: a zahir is "some unobtrusive object" which captures the mind's full attention. "[T]he zahir—so apparently unremarkable but in the event so potent—represents any unexpectedly obsessive image or memory" (328). The three *Zahir*s in which Somebody sails are vehicles for the alteration of time and space.

On the third *Zahir*, reality and fantasy are intermeshed; as with other crucial structures, Barth emphasizes the structural reality of the situation regardless of the fantastic narrative structure. The narrator reluctantly admits that the ship's design is 'quixotic' at best; he and his shipmate are sailing without outside assistance ("without radio, weather service, hydrographic charts" etc.), and when a storm comes up, the narrator remarks that "[w]hatever he imagines, the storm is real" (566–67). Nearly drowned, Somebody instead arrives in modern-day Maryland. Does Somebody achieve his goal and is Barth asserting the 'reality' of transcendent experience? Time and space are altered again for Somebody at the hospital; Baylor tells us that he learns his "true" name and 'crosses over' again, following Death. (Even here, Barth questions the "reality" of the narration and its reliability, giving the reader the choice of deciding whether the narrator is being treated for physical hurts or mental illness.)

Yet Somebody's *other* goal, mentioned before he left Baghdad, was to create a heterotopic space for himself and Yasmīn, and this situation is left unresolved. He initially set sail with Yasmīn to find a new place, explaining to Baghdad friends that:

> Yasmīn's world is truly not his, nor his hers; if we can't intervisit between them, we'll try to reach some third that's neither—an island as far out of her time as out of his, where nothing fixes us in any world except Nature's—and there improvise our own world out of elements from all three. (563)

Whether he achieves this world is open to speculation, depending on whether we consider the story itself as "real." We are told that the storm which wrecked him, and his return to the Maryland shore, are "real" enough. Will he join Yasmīn after death (the ultimate transcendent experience—and possibly a replacement, although unknowable, "reality")? Regardless, Somebody had stated his intent to remain in nature (i.e., the world), indicating that he really does wish to alter reality, not just perception.

A preoccupation with death as the ultimate 'crossing-over'—and Time as the enemy of humanity—dominate two more recent works, *Once Upon a Time* and the short story collection *On With the Story*. In *Once Upon a Time*, a footnote from the author informs us that "vertiginous Time is the very medium that floats this opera."[35] Although time is privileged over space, Barth still grapples occasionally with dynamic space. Most often, Barth discusses movement itself (whether in terms of time or space, whether *in* time or *in* space) as the element connecting the "when" and "where." The couple in *Once Upon a Time*, the "us" sailing aboard the *US*, engage jointly and separately in "kenosis," which the narrator tells us is what "the Greeks called this emptying of the spirit's vessel in preparation for a refill."[36] The couple, sailing on an impromptu quest—which is being recorded by the author-narrator and male half of the partnership—participates literally and metaphorically in life as a process or "journey." The "shape" (or space) of this journey, it's plotted course, involves a continual interaction between past, present, and future events. Neither the vessel itself nor the topography of the journey represent the essential point of their respective metaphors; rather, both center on the *process* of emptying or the process of charting and following a path. Barth turns to physics as well to explain his point, comparing the "continuous improvisation" of storytelling and living life (particularly the activity of giving meaning to life) with concepts such as "self-organized criticality" and "controlled indeterminacy."[37] Our chaotic lives will eventually organize themselves, as a natural outcome of the "chaotic" system—or more specifically, arbitrariness (chaos) and meaning (organization) are not mutually exclusive terms, but compatible ones.

Barth's most integrated use of physics, however, comes in *On With the Story*. Working metaphorically with quantum theory (in which each particle of matter has a wave function), his storytelling couple invents the characters "Amy" and "Richard" who debate the merits of conceptualizing themselves as particles or waves:

> And our selves, those posited centers of gravity that we're in the habit of calling, e.g., Amy and Richard? Waves, definitely: mere ever-changing configurations of memories and characteristics embodied in those other waves, our minds and bodies . . . Does that not imply . . . that, indeed, all human relationships, are waves? . . . even in the stablest of instances the dynamic of their relation is ever in flux, continually disequilibrating and (Let's cross our fingers) continually returning toward equilibrium.[38]

Setting aside the observation that Barth's discussion mixes physics and philosophy, he asserts that individuals (like quantum particles) interact dynamically. In a sense we might define a person through the dynamics of social relationships. The space we live in is necessarily larger than our corporeal bodies because it must make room for dynamic configurations.

With their wave theory of relationships, Amy and Richard challenge conventionality, not least the literary convention that established marriages stagnate and that a crisis (as this couple have experienced) will fragment a 'solid' marriage. Instead, his couple chose to reaffirm their life by conceptualizing it in terms of dynamics. Continual change, in effect, means that a situation that deteriorates can also be reversed (since anything in motion is easier to redirect then anything inert). Of all of Barth's characters, this couple and the narrating couple likewise become transcendent centers of activity and creative force. "Long may *we* wave, then" assert the narrators, in a reaffirmation of "love." Unlike Somebody, who experiences literal space-time transference, Barth's more recent characters reconceptualize their bodies in space; they replace the body-space dichotomy with the paradigm of the wave function. "Waves all, all waves . . . right on out to the exploding universe."[39] Instead of asserting postmodern skepticism, Barth seems to have come full circle to Emersonian idealism.

6
Conclusion

HAS AMERICAN IDEALISM INFECTED AMERICAN POSTMODERN TEXTS? Hardly, if we consider texts from Pynchon, DeLillo, and others whose conspiracy theories leave the reader doubting factual existence, since what one "sees" is attacked as a superficial representation of collusion with a false culture. Despite Jameson's claim, it is easier to find representations of disjointed, fragmented space than more positive visions. Championing the positive side, there is (oddly enough) Jim Jarmusch, an American auteur whose films engage in a dialogue about the nature of American space and its mythic and spiritual dimensions. By drawing a film into the discussion at this point, I do not mean to raise questions which can not be answered in a conclusion—such as the implications of comparing cinema and literature, particularly the various uses of similar terms (author, text, point of view), and the different ways the two modes represent "reality." Jarmusch's film *Dead Man* (1995) presents tempting parallels to the other texts discussed so far. The film also emphasizes an important element in the consideration of American space—that is, how American space continues to be manufactured by myth, particularly the western mythos. In this manner, Jarmusch joins Barth in addressing myth in general (in relation to the United States), and writers like Sam Shepard and Thomas Pynchon who address Western myth in particular.

With a style typical for postmodern works, *Dead Man* features arbitrary events, obscure and detached dialogue, ineffective communication, and cynical and defeatist characters.[1] Additionally, *Dead Man* is in many ways a composite story of the texts mentioned so far in this study, in its reaffirmation of transcendentalism through spatial representation. The central character William Blake (Johnny Depp) arrives in the wilderness of the Western frontier escaping lost love and looking for a job. He soon finds himself dying from a gunshot wound as he tries to complete an impromptu spiritual "vision" quest. Like Kumminz, Blake is guided through a hellish landscape by a Virgil-like Native American called Nobody (Gary Farmer). As with the 'journeys' of Thoreau, C, and Somebody, Blake's journey involves the concept of the "way station," represented in architectural terms. His journey is framed by 'realistic' representations of social space. Blake exits his train at "Machine," a hellish, Dickensian town (as is evident not only from the dialogue and shots of the factory but from its owner's name, Dickinson). He leaves, traveling through the wilderness by horse and ends his journey at a North Pacific Indian village. This village, a symbol of his renewed ('native') spirituality, is the departure point for his (nearly) dead self. He drifts across the water into the "other realm."

Like Barth, Jarmusch creates "real" spaces out of myth and uses realism mythically, thoroughly confounding the two ontologies. When William Blake exits the train and walks through the town of Machine, we are struck by the realism—actually, the superrealism—in the film's exaggeration of the dusty, clapboard buildings, the peeling paint, the wooden walkways cluttered with junk, the unhewn tree branches holding up tin roofs. In a separate shot, we see how paint was worn from a building where mud and water had splashed up, the material decay emphasized by an abandoned coal stove set in front of the wall. Then visual realism is taken to extremes, the site seamlessly crossing over into unrealistic, mythic "hell." We see coffins being manufactured along the street, piles of cow bones and human skulls, a horse urinating, a woman engaging in an oral sex act with a man . . . crude sights of dusty, unsmiling people and images of death. To conclude this montage, in a long shot of the street, we see the factory at the end with its billowing smoke. The camera leads us through its riveted, iron gate (an invocation of Dante's gates of hell). The town's realism, which dupes the viewer and temporarily validates the myths of the Western town, turns realism into fantasy by invoking a competing (while complimentary) image of the mythic realm of the Underworld. The realistic setting of the Western town, once exaggerated, comments on the ideological underpinnings of this cultural construct. As a special, heterotopic site and as the broader, social setting for the community, it "reflects and speaks about" itself.

At the other end of Blake's journey, a similar, realistic treatment of the native village achieves the opposite goal. Its realism serves to comment not on the façade of realistic representation (as with the Western town) but on the vague, under-detailed mythos of Native American religion. As Blake and Nobody approach and then enter the village, we see the large, wooden palisade surrounding the homes. The exterior and interior are unpainted except for a few large designs in primary colors. The leather-covered entrance leads to wide open dirt lanes edged by housing built against the wall. Large totems and tall poles with carvings at top are scattered around the complex. As in the frontier town, people doing outdoor work (drying skins, cooking food) sit in plain sight along the edges of the lane. The entire scene appears to have been based on archaeological research; an abandoned sewing machine sitting beside scattered bones reminds us of the "reality" of native life at this time, with its invasions of European-based technological society. This village represents itself as a detailed, cultural "whole" with an intact, functioning system regulating daily life. Unlike the Western town, which is trapped in its own mythologizing, this representation of a native village confronts romanticized visions with ethnic realism.

Because of the juxtaposition of realism and myth, *neither* site successfully represents accepted definitions of reality or myth. Although Blake passes through both sites, he acts as a passive receptor of the revisioning of space taking place in the film. He is the receptor as well of a "vision quest" which happens to him through a series of accidental events ("fate"). In consideration of his passivity, can he be a transcendental agent? Does Blake recreate space? He's hurt and appears only mildly interested in his own transformation; he seems to have little control over events. The narrative convention of a central perspective is dominant in the film; the dynamics of spatial sites are viewed through the camera eye. The film itself re-imagines spatial configurations and representations; like Blake, the viewer is a passive receptor of scenic information. Still, the narrative "center" becomes diffused by large landscapes and 'placeless' places. Topography is dislocated from cartography, and realistic landscape and architecture sit oddly in a dislocated, mythologized area. Even the full control of authorship represented by the camera eye cannot account fully for the dynamics of the transcendental experience. As with the last "translation" of Somebody the Sailor, the process of transcending space displaces fixed definitions of mythic versus realistic space.

The reality of space, which we rely on as a basis for social organization, is unfixed when coupled with transcendence, transforming the nature and purpose of heterotopic sites in the process. At bottom, the search (or quest, in Blake's case) does not even have an object or an agent. Why is Blake dying in the wilderness? Because of his lost love? Because he was tricked arbitrarily into murder? Because he is essentially homeless, lacking the security of an established social net and predictable social mores of the Eastern United States? Blake did not even participate in creating the diametrically-opposed Western landscape which we see; he arrived in it, much as Somebody spontaneously arrives in Baghdad. The individual as a perceptual center is less important than the individual as part of a moving system. Participation within that field, as Blake discovers, alters reality. Human beings, as Lefebvre states, "act and situate themselves in space as active participants . . . in a series of enveloping levels each of which implies the others."[2] However, the activity of the site(s) themselves involves even an unwilling participant like Blake. The elements of the field act and react in a mutually compelling series of events.

The search for truth is a traditional paradigm that presupposes an agent searching purposefully for a real object, "the truth." This paradigm remains a powerful construct, as can be seen in the popular TV series *The X Files*, where Fox Mulder repeatedly invokes his search for "the Truth." He is an active agent engaging mysterious forces in a quest for the ill-defined concept of "Truth." Two constructs are at work in this show since the agent on a quest is battling a field of forces which pulls him into nebulous schemes. These constructs offer critiques on each other, so that neither would gain prominence except for the audience's identification with the heroic agent and his search. Foucault, however, dismisses the paradigm of the search as an outdated, transcendental, nineteenth-century habit of thought. The nature of the search relies on the assumption that although purposeful, a search for origins can never be concluded. In *The Archaeology of Knowledge*, Foucault explains that searchers hope that "it never be possible to assign, in the order of discourse, the irruption of a real event; that beyond any apparent beginning, there is always a secret origin."[3] The search consequently leads from point to point, infinitely, without discovering a "real event" but maintaining faith in an eventful origination.

For Foucault reality rests not merely in phenomenology (a system of objects that again could lead us on the search for origins) but in the uniqueness of any event. Therefore a discursive event (any type of communication in an intellectual discipline) does not signify something else, such as an origin or an author's intention. Its uniqueness relies on its present condition: "To reveal in all its purity the space in which discursive events are deployed is not to undertake to re-establish it in an isolation that nothing could overcome; it is not to close it upon itself; it is to leave oneself free to describe the interplay of relations within it and outside it."[4] Despite Foucault's insistence that he establish "*rules of formation*," his description of how we should articulate a discursive event emphasizes free play.[5] Although hoping to replace transcendentalism with more logical rules, he nevertheless offers us a poetic vision.

Foucault battles reductive structuralism with play, and his supposition of freedom for the perceiver / describer implies that we can transcend our singular perception in a field to encompass the field's multiplicity. Foucault's description of a new discursive "field" of discourse events parallels current architectural understanding of spatial fields.[6] Architects have also expressed interest in the "interplay of relations" in a spatial field—the patterns of movement, the relationships between sites, and the dynamics of natural and cultural forces in that field (for example, with the conjunction of media and architectural surfaces). The remaining unknown factor—the organizing force or creator of a field—is replaced by the function of interconnections. Given the arbitrariness of boundaries and containers, a field—as described by chaos theory—is a self-organizing system.

Even in architectural theory, Transcendentalism remains the rhetorical enemy and is still set in opposition to dynamic reality. Stephen Perrella's definition of hypersurface architecture as "altered" surface depends on this binary opposition, his discussion of which he entitles "antiTranscendental defections."[7] According to him, "any process that assumes an ideality as an ultimate end is doomed to failure," and we should just "accept" our "schizophrenic condition" instead of "attempting to . . . transcend it."[8] Space is no longer—if it ever was—satisfactorily divided into the "real" and "ideal." Our cultural schizophrenia may come from the remnants of Western philosophy—or it may come from our inability to integrate the real and ideal. In either case, Perrella and his colleagues' descriptions of dynamic space and their attempts to involve architecture in this dynamism duplicate Thoreau's vision of transcendence.

In associating play and freedom in a spatial field with transcendence, I am following one strain in Transcendentalism; Foucault bases his objections to transcendental thought on another strain, one which is admittedly more historically prominent. Foucault intends his programmatic use of "discontinuity, rupture, threshold, limit, series, and transformation" to "free the history of thought from its subjection to transcendence . . . that circle of the lost origin."[9] He opposes historical positivism, the search for origins, universal forms, and the foundation of being, with "the difference of our present" and "an analysis of practice" of the subject itself, not its supposed origin.[10]

The Transcendentalists were involved in a search for origins, offering the possibility for direct experience of God unmediated by church or history. Emerson hoped to achieve a transcendent connection with the Oversoul (his conception of the origin); Thoreau, too, searched for origins not by looking up but by looking down, discovering his spiritual foundations in the ground and in the roots of organic matter (and language as well). In *Nature*, Emerson describes the object of the search—God's "higher law"—as a "rule" which "holds true throughout nature. . . . this Unity . . . is like a great circle on a sphere, comprising all possible circles; which, however, may be drawn and comprise it in like manner" (40–41). Emerson's great circle sounds remarkably similar to Jorge Luis Borges' total book; each circle (or book) presents one perspective out of an unlimited number of perspectives of God (or the central book).

Emerson touts the idea of unity despite his own contradictory statements. (His essay "Circles" posits a different image, in which spiritual growth expands outward in unlimited concentric circles.) He also disliked the popular misconception of Transcendentalism as a sterile philosophy. Unlike the practical materialist or the self-centered idealist, the true Transcendentalist (in his essay on the subject) "adopts the whole connection of spiritual doctrine. He believes in miracle, in the perpetual openness of the human mind to new influx of light and power . . ." (195). The goal of the Transcendentalist is "fuller union with the surrounding system" (206). Emerson remains ambiguous on whether this system operates as a closed unity or as an open process.

Foucault primarily objects to Transcendentalism as a discourse of origins. Yet out of Emerson and Thoreau's transcendentalism emerges an understanding of interconnected and kinetic space. Both men demonstrate their fascination with spatial fluidity as a correlative of spiritual freedom. Emerson celebrates the "fluid and volatile" nature of the universe (168). For Thoreau, "the very globe continually transcends and translates itself, and becomes winged in its orbit" (567). This continual translation denies the need for a center; the translation itself serves as a unifying force.

We can begin to see a tenuous but consistent line of thought meandering from Emerson and Thoreau to the Futurists and Cummings, through Borges to Barth and Lefebvre. Kinetic activity permeates concepts of space and time, material and spiritual being, social and natural constructs, in opposition to rigid systems, laws, and logisms. Sequential and mobile forms replace binarism, linearity, and progression. In "The Literature of Replenishment," Barth traces this metamorphosis to the modernists who, "carrying the torch of romanticism, taught us that linearity, rationality, consciousness, cause and effect, naive illusionism . . . are not the whole story."[11] Borges' narrator of "The Library of Babel," although a skeptical thinker, tentatively proposes an alternative to the single unity—infinite, periodic disorder. The periodic repetition of a limited chaos "would constitute an order: Order itself" (88). Borges anticipated the general theoretical stance of late-twentieth-century physics, that chaotic systems create their own order. In this sense, a system formulates itself without an originating force, an idea particularly attractive to Barth.

Barth, who has adopted chaos theory as a "paradigm" of "the way many people live" and possibly of "human life in general," finds the concept of self-organized criticality in weak chaos most useful.[12] Citing a January 1991 *Scientific American* article, Barth explains that self-organized criticality is a model of "dynamic systems." Accordingly, "chain reactions of all sizes are an integral part of [their] dynamics" so that instead of reaching equilibrium, complex systems continuously "evolve from one metastable state to the next." Systems consist of chain events, self-generated and "on the border of chaos."[13] The threat of entropy destroying a system or nonlinearity dissolving into chaos gets turned on its head, and dynamism ironically provides stability.

Similarly, Lefebvre proposes multiple centers of activity in place of systematic unity, centers which serve to organize reality into webs of relationships whose fixed points of origin and authorship are replaced by activity itself. "There is no 'reality' without a concentration of energy, without a focus or core . . . What is the 'subject'? A momentary centre. The 'object'? Likewise. The body? A focusing of active (productive) energies. The city? The urban sphere? Ditto."[14] In Lefebvre's schematic of space, movement takes a central, pervasive role, one that reunites the split between time and space. Although continual change seems to threaten stability, change itself involves all parts of the system and holds those parts together.

"Any centrality," argues Lefebvre, "once established, is destined to suffer dispersal, to dissolve or to explode from the effects of saturation, attrition, outside aggressions, and so on. This means that the 'real' can never become completely fixed, that it is constantly in a state of mobilization."[15] In postmodern thought, reality comes unfixed and destruction merely anticipates a new organization. The process of reorganization remains a constant factor. In E. E. Cummings' terminology, motion equals aliveness, and the verb "is" animates being. His focus on the verb of being anticipates Heidegger's notion of activity and purpose as motivating everyday life and its objects within a field of interactions. Both focus not on objects themselves but on the process of interaction. According to Charles Guignon, in Heidegger's image of a workshop (in *Being and Time*) "what shows up for us initially is the hammering," the activity that connects the person and the tools, which are secondary and dictated by whatever "project" the worker has underway. The need to hammer brings this object into focus, rather than the sight of the hammer suggesting a compulsion to hammer.[16] "Hammering" defines the space, and syntactically, the verb is emblematic of being.

Lefebvre criticizes Heidegger as being too involved with the concepts of time and history, but Heidegger's limitation may rest instead with his conception of the whole, and his validation of the hermeneutic circle as a paradigm of human understanding. "[T]he parts cannot be interpreted without an understanding of the whole, but the whole cannot be grasped without an understanding of the parts." Although "avoid[ing] the traditional model of the subject as knower standing over against what is to be known," Heidegger still "posits a whole" in his insistence that self and world form a circle.[17] Obliquely, Heidegger asserts the figure of the container into the paradigm and brings us full circle to the question of origins. How do we describe the "whole" of this circle (or, to use earlier metaphors, the sphere, the library, the Unity) if the circle is dynamic, in flux, and can be 'viewed' and described from multiple points of view? In this sense, 'meaning' is elusive.[18]

Barth's discussion of chaos theory in "4 1/2 Lectures" reminds us that the scale of measurement momentarily determines the container. A map of the East Coast looks entirely different if we apply fractal geometry; we look at the coastline as if with a microscope of increasing power and find infinitely more detail in each mile. If we fly by jet above the cloud line, we discover a new field of space where a floor of clouds covers a huge distance and places the sun over a larger horizon. As we travel, moving through spatial fields, and as we adjust our perception of these fields, we participate in the organization of space. Emerson is perhaps closer than Heidegger in the philosophical conceptualization of our self-centered views of a destabilized, unobjective universe. In a non-Western paradigm, Emerson forgoes the division between subject and object in favor of a completely subjective world. The idealist sees "the procession of facts you call the world, as flowing perpetually outward from an invisible, unsounded center in himself, center alike of him and of them, and necessitating him to regard all things as having a subjective or relative existence . . ." (195).

In the dialogue about the role of the individual subject, a difficulty arises concerning how to talk about the self and its "central" viewpoint without either elevating or dismissing the perceptive self. In considering social space, Lefebvre laments the impoverishment of people labeled as "inhabitants" of space.[19] In contrast to the more exalted figure of architect and urban planner, the person who "uses" space is marginalized, because the day-to-day occupation of space lacks commercial value (after the initial exchange of money for space). For the individual, however, the constant participation in spatial fields remains purposeful and real. Lefebvre sees this need most clearly in vacation spots, where " 'people' demand a qualitative space. The qualities they seek have names: sun, snow, sea. . . . Neither spectacle nor mere signs are acceptable. What is wanted is materiality and naturalness as such. . . .[20] Thoreau would understand qualitative space, a space that is concrete, real, and rooted in existence.

Foucault also recognizes the primacy of the "real" in his definition of heterotopias. In tracing the history of space, Foucault describes modernity as concerned with "external space": "Our epoch is one in which space takes for us the form of relations among sites." A site in turn is "defined by relations of proximity between points or elements."[21] One feature of qualitative space rests in the nature of relationships that occur in discrete places; these relationships also occur across scale, connecting and reconfiguring identifiable fields of space in unlimited ways. The dynamic nature of any site belies a fixed center; rather, each point in the structure can be taken as a point of perspective. This perspective center, as Lefebvre argues, shifts as perceptions change. Action itself replaces centrality as the crucial mode of existence.

Notes

Preface

1. See Charles Olson, *Call Me Ishmael: A Study of Melville* (New York: Grove, 1947).

2. Edward Soja, *Postmodern Geographies: The Reassertion of Space in Critical Social Theory* (New York: Verso, 1989), 11.

3. Michel Foucault, "Of Other Spaces," trans. Jay Miskowiec, *Diacritics: a Review of Contemporary Criticism* 16, no.1 (spring 1986): 24.

4. Joan Burbick, *Thoreau's Alternative History: Changing Perspectives on Nature, Culture, and Language* (Philadelphia: Univ. of Pennsylvania Press, 1987), 9.

5. Unless otherwise stated, quotations from the prose of Henry David Thoreau are from *Henry David Thoreau*, ed. Robert F. Sayre (New York: Library of America, 1985), 400

6. E. E. Cummings, *The Enormous Room*, ed. George Firmage (New York: Liveright, 1978), 143.

7. Ada Louise Huxtable, "The New Architecture," *The New York Review of Books* (April 6, 1995): 18–19.

Chapter 1. Introduction: Architectural Space

1. Edgar Allan Poe, *The Complete Tales and Poems of Edgar Allan Poe* (New York: Vintage, 1975), 248.

2. Andrew Marvell, *Andrew Marvell* (Oxford: Oxford Univ. Press, 1990), 18–19.

3. Fredric Jameson, *Postmodernism, or, The Cultural Logic of Late Capitalism* (Durham: Duke Univ. Press, 1995): 415.

4. Ralph Waldo Emerson, *Essays: First and Second Series*, ed. Joel Porte (New York: Vintage, 1990), 157–58.

5. Ralph Waldo Emerson, *Selections from Ralph Waldo Emerson*, ed. Stephen E. Whicher (Boston: Houghton, 1957), 24. Unless otherwise stated, quotations of Emerson's prose are from this collection.

6. Richard Poirer, *A World Elsewhere: The Place of Style in American Literature*, 2d ed. (Univ. of Wisconsin Press, 1985), 17.

7. Gertrude Stein, *Tender Buttons: Objects, Food, Rooms* (New York: Claire Marie, 1914), 66. All other page references to this text will be noted by internal citation.

8. Soja, *Postmodern Geographies*, 11.

9. See the essays included in *Designing for Human Behavior: Architecture*

and the Behavioral Sciences, Jon Lang, Charles Burnette, Wlater Moleski, and David Vachon, eds. (Stroudsburg, Pa.: Dowden, Hutchinson and Ross, 1974). See also Michael Pearson and Colin Richards, "Ordering the World: Perceptions of Architecture, Space, and Time," in *Architecture and Order: Approaches to Social Space* (New York: Routledge, 1994).

10. Lang, "Theories of Perception and 'Formal' Design," *Designing for Human Behavior,* 108.

11. Brian Massumi, "Sensing the Virtual, Building the Insensible," *Architectural Design: Hypersurface Architecture* 68, no. 5/6 (May–June 1998): 22.

12. Bruno Zevi, *Architecture as Space: How to Look at Architecture*, trans. Milton Gendel, ed. Joseph A. Barry, rev. ed. (New York: Horizon, 1974), 27. Lang critiques the reliance of basic design on the point, line, and plane. He explains that curve, shadow, texture, obliqueness, and other elements important to design go unnoticed when an architect focuses on the principles of basic design. For an explanation and critique of basic design, see Lang, "Theories of Perception and 'Formal' Design," *Designing for Human Behavior,* 98–110.

13. Cornelis Van de Ven, *Space in Architecture: The Evolution of a New Idea in the Theory and History of the Modern Movements*, 3rd ed. (Assen, Netherlands: Van Gorcum, 1987), 30, 38–39.

14. Massumi, *Architectural Design*: 21.

15. Ibid., 22.

16. Bruno Zevi describes the Greek temple as sculptural not architectural, because no one can experience the inside. All rituals took place on the porch. Zevi, *Architecture as Space*, 76–77.

17. For a discussion and examples of cognitive mapping and architecture, see David Stea, "Architecture in the Head: Cognitive Mapping," in *Designing for Human Behavior: Architecture and the Behavioral Sciences,* Jon Lang, Charles Burnette, Wlater Moleski, and David Vachon, eds. (Stroudsburg, Pa.: Dowden, Hutchinson and Ross, 1974), 157–68.

18. Pearson and Richards, "Ordering the World: Perceptions of Architecture, Space and Time," *Architecture and Order,* 3–4.

19. Henri Lefebvre, *The Production of Space,* trans. Donald Nicholson-Smith (Cambridge: Blackwell, 1991), 170.

20. Ibid., 83.

21. Charles Olson, *Call Me Ishmael,* 25.

22. Zevi, *Architecture as Space*, 187.

23. Cited in Stephen Kern, *The Culture of Time and Space: 1880–1918* (Cambridge: Harvard Univ. Press, 1983), 145.

24. Milton Cohen, *PoetandPainter: The Aesthetics of E. E. Cummings's Early Work* (Detroit: Wayne State Univ. Press, 1987), 119.

25. E. E. Cummings, *Complete Poems, 1904–1962*, ed. George Firmage, rev. ed. (New York: Liveright, 1991), 122. Unless otherwise stated, all quotations of poetry by E. E. Cummings come from this volume.

26. Foucault, "Of Other Spaces," 24.

27. Ibid., 24–25.

28. Ibid., 27.

29. Ibid., 23.

30. Henry James, "The Jolly Corner," *The Novels and Tales of Henry James,* vol. 17 (New York: Scribner, 1937), 470, 453.

31. See Steven Pinker, *The Language Instinct* (New York: William Morrow, 1994).

32. Fredrick Jameson, *The Prison-House of Language: A Critical Account of Structuralism and Russian Formalism* (Princeton: Princeton Univ. Press, 1972), 37.

33. Richard Harland, in *Beyond Superstructuralism: The Syntagmatic Side of Language* (New York: Routledge, 1993), favors the syntagmatic acts of combination and synthesis which occur within a sentence over the more limited paradigmatic creation of meaning out of likeness and contrast. While I agree that the context of grammatical forms as a sentence defines individual words, I question whether this context is a "domination of syntagmatic wholes" (54). Rather, as Harland himself points out, the reader "juggles" parts and wholes in order to create meaning (23). In this sense the syntagmatic and paradigmatic both contribute to the dynamics of the text. In a way, while "grammatical categories . . . recompose a new three-dimensionality in sentences," the consideration of both vertical and horizontal functions of text creates a sort of fourth-dimensionality (18).

34. Zevi, *Architecture as Space*, 28.

35. Gérard Genette, *Figures of Literary Discourse*, trans. Alan Sheridan (New York: Columbia Univ. Press, 1982), 50.

36. Gaston Bachelard, *The Poetics of Space,* ed. and trans. Maria Jolas (Boston: Beacon, 1994), xvi, xxiii.

37. Ibid., xx–xxi.

38. Ibid., 6.

39. Ibid., 18.

40. Nathaniel Hawthorne, "The Old Manse," *Mosses From an Old Manse* (Ohio State Univ. Press, 1974), 17. All other quotations of Hawthorne's prose refer to this edition.

41. Bachelard, *The Poetics of Space,* xxi, xxiii.

42. Clare Cooper, "The House as Symbol of the Self," *Designing for Human Behavior,* 131.

43. Cummings, *The Enormous Room*, 90.

44. Marilyn Chandler, *Dwelling in the Text: Houses in American Fiction* (Berkeley and Los Angeles: Univ. of California Press, 1991) 23.

45. Stephen Perrella, "Hypersurface Theory: Architecture><Culture," *Architectural Design*: 10.

46. Ibid., 11.

47. Ibid., 12.

48. Martin Heidegger, "Building Dwelling Thinking," *Poetry, Language, Thought*, trans. Albert Hofstadter (New York: Harper, 1971), 156.

49. Ibid.

50. Ibid., 154.

51. Albert Einstein, foreword to *Concepts of Space: the History of Theories of Space in Physics,* by Max Jammer, 3rd ed. (New York: Dover, 1993), xvii.

52. Jammer, *Concepts of Space,* 216. For a complete overview of this debate, see chapter 6.

53. Brian McHale, *Postmodernist Fiction* (New York: Methuen, 1987), 39 and Perrella, *Architectural Design*: 11.

Chapter 2. Transcending Space: *Walden*

1. Octavius Brooks Frothingham, *Transcendentalism in New England* (New York: Harper, 1959), 238.
2. Ibid., 20.
3. Catherine Albanese, in *Corresponding Motion: Transcendental Religion and the New America* (Philadelphia: Temple Univ. Press, 1977) discusses the kinetic language and metaphors of American transcendentalism.
4. Cited in Robert Richardson Jr.'s *Henry Thoreau: A Life of the Mind* (Berkeley and Los Angeles: Univ. of California Press, 1986), 345.
5. Cited in Richardson, *Henry Thoreau*, 94.
6. Frothingham, *Transcendentalism*, 18.
7. Kern, *The Culture of Time and Space*, 132.
8. Frothingham, *Transcendentalism*, 12.
9. Ibid., 140.
10. Emerson, *Essays*, 187.
11. Ibid., 187.
12. Albanese, *Corresponding Motion*, 29.
13. Ibid., 158.
14. Van de Ven summarizes these concepts in *Space in Architecture*, 46.
15. Van de Ven, *Space in Architecture*, 27.
16. Burbick, *Thoreau's Alternative History*, 65, 69.
17. Ibid., 60–64.
18. Ibid., 74, 31.
19. Foucault, "Of Other Spaces," 22.
20. Soja, *Postmodern Geographies*, 11.
21. Cited in Judith Fryer, *Felicitous Space* (Chapel Hill: Univ. of North Carolina Press, 1986), 48–49.
22. Burbick, *Thoreau's Alternative History*, 45, 46.
23. Thoreau, *The Journal*, vol. 2, 253.
24. Cited in Kern, *The Culture of Time and Space*, 135.
25. Ibid., 135.
26. Thoreau, *The Journal*, vol. 4, 152.
27. Heidegger, "Building Dwelling Thinking," 152.
28. Lang, "Theories of Perception," 100.
29. Lefebvre, *The Production of Space*, 191.
30. Thoreau, *The Journal*, vol. 3, 154.
31. Thoreau, *The Journal*, vol. 7, 118–19.
32. Thoreau, *The Journal*, vol. 3, 182.
33. Michel Foucault, *Power / Knowledge: Selected Interviews and Other Writings, 1972–1977*, ed. Colin Gordon, trans. Colin Gordon et. al. (New York: Pantheon, 1980), 149.
34. Bachelard, *The Poetics of Space*, 11–12.
35. Joseph Allen Boone, "Delving and Diving for Truth: Breaking Through to

Bottom in Thoreau's *Walden*," *Critical Essays on Henry David Thoreau's Walden*, ed. Joel Myerson (Boston: G. K. Hall, 1988): 164.

36. Cited in Kern, *The Culture of Time and Space*, 137.

37. Ibid.

38. Lefebvre, *The Production of Space*, 229.

39. Kern, *The Culture of Time and Space*, 136.

40. Bachelard, *The Poetics of Space*, 12.

41. Albanese, *Corresponding Motion*, 60.

42. Lefebvre, *The Production of Space*, 173.

43. Ibid., 173–74.

Chapter 3. Dynamic Space: *The Enormous Room*

1. Unless otherwise stated, in this chapter the prose of E. E. Cummings is quoted from *The Enormous Room*, ed. George Firmage (New York: Liveright, 1978), 3.

2. Foucault, "Of Other Spaces," 24.

3. Lefebvre, *The Production of Space*, 369.

4. Ibid., 23.

5. Lang, "Theories of Perception," 100.

6. Ibid., 108.

7. Kent Bloomer and Charles Moore, *Body, Memory, and Architecture* (New Haven: Yale Univ. Press, 1977), 34–35.

8. In citations from Cummings' work where material is omitted, ellipses are set off with brackets to differentiate my own ellipses from the author's. Cummings rarely uses brackets. Cummings, *The Enormous Room*, 143.

9. E. E. Cummings, "The New Art," *E. E. Cummings: A Miscellany Revised*, ed. George Firmage (New York: October House, 1965), 10–11.

10. Ibid., 10, 5.

11. Ibid., 10. Cited in "The New Art."

12. E. E. Cummings often demonstrates a marked lack of sensitivity. While he loved to defend downtrodden people—as he did when loyally supporting Ezra Pound during his trial for treason—Cummings also loved to offend his audience (and often offended while defending). Following World War II, Cummings still made use of the term "kike" in his poem "a kike is the most dangerous" and caused many readers to label him anti-Semitic. Because *The Enormous Room* celebrates the "strengths" of so many minorities it is difficult to separate the intentional from the unintentional prejudices and stereotypes. The reader can find obvious examples of both.

13. Christopher Mott, "The Cummings Line on Race," *Spring, A Journal of the E. E. Cummings Society*, New Series, no. 4 (October 1995): 75.

14. Ibid.

15. Foucault, "Of Other Spaces," 24.

16. Lefebvre, *The Production of Space*, 39.

17. Foucault, "Of Other Spaces," 26.

18. Arthur Eddy, *Cubists and Post-Impressionism* (Chicago: A. C. Clurg, 1914), 47.

19. Umberto Boccioni, "The Exhibitors to the Public 1912," *Futurist Manifes-*

tos, ed. Umbro Apollonio, trans. Robert Brain et al. (London: Thames and Hudson, 1973), 50.

20. E. E. Cummings, *i six nonlectures*, (Cambridge: Harvard Univ. Press, 1959), 53.

21. Quoted in Cohen, *PoetandPainter,* 186.

22. Cummings, *Complete Poems,* 221.

23. Filippo Tommaso Marinetti, "Technical Manifesto of Futurist Literature (1912)," *Marinetti: Selected Writings*, ed. R. W. Flint, trans. R. W. Flint and Arthur Coppotelli (New York: Farrar, Straus and Giroux, 1971), 86.

24. Ibid., 88–89, 85.

25. Quoted in Cohen, *PoetandPainter*, 186.

26. Roman Jakobsen, "Closing Statement: Linguistics and Poetics," in *Style in Language*, ed. Thomas A. Sebeok (New York: Technology Press of Massachusetts, 1960), 350.

27. Umberto Boccioni, "Plastic Dynamism 1913," in *Futurist Manifestos*, 92.

28. Ibid., 94.

29. Quoted in Cohen, *PoetandPainter*, 153.

30. Stea, "Architecture in the Head," 157.

31. Ibid., 163.

Chapter 4: Inverted Space: *EIMI*

All quotations of the prose and poetry of E. E. Cummings in this chapter come from *EIMI*, 2d ed. (New York: Sloan, 1933), unless otherwise noted.

1. For other details of this trip see Richard Kennedy, *E. E. Cummings revisited* (New York: Twayne, 1994), 85.

2. Cummings gave these nicknames to Mr. and Mrs. Charles Malamuth. Malamuth was a foreign correspondent; his wife was the daughter of Jack London. See Kennedy, *E. E. Cummings revisited*, 85.

3. E. E. Cummings, introduction to *The Enormous Room,* 3rd ed. (New York: Random, 1949), viii.

4. Norman Friedman, *e.e. cummings: The Growth of a Writer*, (Carbondale: Southern Illinois Univ. Press, 1964), 11.

5. Cummings uses the term "unhero" again in his *i six nonlectures*.

6. Lefebvre, *The Production of Space,* 294.

7. Roshi Philip Kapleau, *The Three Pillars of Zen: Teaching, Practice, and Enlightenment*, rev. ed. (New York: Doubleday, 1989), 70.

8. Cited in Kapleau, *The Three Pillars of Zen*, 71.

9. Lista 18.

10. As mentioned in the previous chapter, Cummings studied the Futurists' manifestos included in Arthur Eddy's 1914 summary of contemporary art movements, *Cubists and Post-Impressionism*.

11. Lefebvre, *The Production of Space*, 292.

12. Robert Hershberger, "Predicting the Meaning of Architecture," in *Designing for Human Behavior,* 149–50.

13. Ibid., 152.

14. Ibid., 152, 151.

15. For a history of Cummings' romantic vision and his celebration of the imag-

ination, see Friedman, *Growth of a Writer*, particularly the introduction. For a specific treatment of *EIMI*, see chapter 7.

16. In his early poems Cummings often speaks cynically of (among other criticisms of religion) the religious institution's inability to deal with human sexuality ("god gloats upon Her stunning flesh. Upon") and its usurpation of materialist values ("the season 'tis, my lovely lambs," which continues "of Sumner Volstead Christ and Co.").

Chapter 5. Controlling / Creating Space: *Lost in the Funhouse*

1. John Barth, *Lost in the Funhouse: Fiction for Print, Tape, Live Voice*, rev. ed. (New York: Anchor, 1988), 169. Unless otherwise indicated, quotations of Barth's prose come from this text.

2. Lefebvre, *The Production of Space*, 294.

3. As Jameson suggests, "cognitive mapping" is a misnomer. The powerful image of the map overwhelms the total idea, which identifies the complex mental process of our spatial imaginations. It is, however, a starting point, if one considers that anything can be described within an infinite series of maps. See Jameson, *Postmodernism*, 409.

4. Lynn Stewart, "Bodies, Visions, and Spatial Politics: A Review Essay on Henri Lefebvre's *The Production of Space*" *Environment and Planning D: Society and Space* 13, no. 5 (October 1995): 615.

5. Thomas Carmichael, "Buffalo / Baltimore, Athens / Dallas: John Barth, Don DeLillo and the Cities of Postmodernism," *Canadian Review of American Studies* 22, no. 2 (fall 1991): 244–48. The author explains that the city in *Letters* is a "double-coded" site of anxiety and illumination, yet does not explain why or how Ambrose achieves the latter state.

6. John Barth, *The Floating Opera*, rev. ed. (New York: Doubleday, 1967), 7.

7. Jorge Luis Borges, *Labyrinths*, enl. ed. (New York: New Directions, 1964), 113. All citations of Borges' prose are from this text.

8. "[T]he naive character of the authorial intrusions . . . point to an early attempt by an adolescent author." Zack Bowen, *A Reader's Guide to John Barth* (New York: Greenwood, 1994), 55.

9. Stephen Perrella and Rebecca Carpenter, "The Möbius House Study," *Architectural Design: Hypersurface Architecture* 68, no. 5/6 (May–June 1998): 80–83.

10. Cited in Michio Kaku, *Hyperspace* (New York: Oxford Univ. Press, 1994), 77–78.

11. Tony Tanner, *The City of Words: American Fiction, 1950–1970* (New York: Harper and Row, 1971), 254.

12. Jack Tharpe, *John Barth: The Comic Sublimity of Paradox* (Carbondale: Southern Illinois Univ. Press, 1974).

13. Steven Bell, "Literature, Self-Consciousness, and Writing: The Example of Barth's *Lost in the Funhouse*," *The International Fiction Review* 11, no. 2 (summer 1984): 93.

14. E. P. Walkiewicz, *John Barth* (New York: Twayne, 1986), 89.

15. Kaku, *Hyperspace*, 60–61.

16. Cited in John Kasson, *Amusing the Million: Coney Island at the Turn of the Century* (New York: Hill and Wang, 1978), 59.
17. Foucault, "Of Other Spaces," 27.
18. George and Suzanne Hurley, *Ocean City: A Pictorial History* (Virginia Beach: Donning, 1979), 41.
19. Mary Corddry, *City on the Sand: Ocean City, Maryland, and the People Who Built It* (Centreville, Maryland: Tidewater, 1991), 38.
20. Ibid., 30.
21. Kasson, *Amusing the Million*, 65.
22. Ibid., 59.
23. Corddry, *City on the Sand*, 180.
24. Ibid.
25. Foucault, "Of Other Spaces," 27.
26. Russel Nye, "Eight Ways of Looking at an Amusement Park," *Journal of Popular Culture* 15, no. 1 (summer 1981): 66.
27. Ibid., 73.
28. Walkiewicz, *John Barth*, 98.
29. Ross King, *Emancipating Space: Geography, Architecture, and Urban Design* (New York: Guilford, 1996), 123.
30. Ibid., 123.
31. John Barth, *Chimera* (New York: Fawcett Crest, 1972), 75. The following quotations, until otherwise noted, are from this text and page numbers are referenced parenthetically.
32. Lefebvre, *The Production of Space*, 361.
33. Foucault, "Of Other Spaces," 24.
34. John Barth, *The Last Voyage of Somebody the Sailor* (Boston: Little, Brown, 1991), 10–11. 154. Until otherwise noted, the following quotations of Barth's prose are from this text and page numbers are referenced parenthetically.
35. John Barth, *Once Upon a Time: a Floating Opera* (New York: Little, Brown, 1994), 131.
36. Ibid., 13.
37. Ibid., 20.
38. John Barth, *On With the Story* (New York: Brown, Little, 1996), 143.
39. Ibid., 144.

Chapter 6. Conclusion

1. Jarmusch has been called a "minimalist" more than a "postmodernist." Only in this latest film has he been accused of turning away from his "implicit humanism" to adopt a "Tarantinoesque flippancy." Despite any "humanist" leanings, his films deconstruct traditional narration in postmodern fashion—substituting other organizational rationales (the circularity of *Night on Earth* for instance) and featuring chaotic and 'meaningless' actions (as in Willie and Eddie's sudden decision to visit Florida in *Stranger than Paradise*). See Ben Thompson, review of *Dead Man, Sight and Sound* 6, no. 7 (July 1996): 41–42. See also Peter Keogh, "Home and Away," *Sight and Sound* 2, no. 4 (August 1992): 8–9.
2. Lefebvre, *The Production of Space*, 294.

3. Michel Foucault, *The Archaeology of Knowledge*, trans. A. M. Sheridan Smith (New York: Pantheon, 1972), 25.
4. Ibid., 29.
5. Ibid., 38.
6. Ibid., 27.
7. Perrella, "Hypersurface Theory," 8.
8. Ibid., 8–9.
9. Foucault, *The Archaeology*, 203.
10. Ibid., 204.
11. John Barth, "The Literature of Replenishment: Postmodernist Fiction," *The Atlantic* 245 (January 1980): 70.
12. John Barth, *Further Fridays: Essays, Lectures, and Other Nonfiction, 1984–94* (New York: Little, 1995), 339.
13. Barth in his "4 1/2 Lectures" in *Further Fridays* cites authors Per Bak and Kan Chen, 338–39. In these lectures, Barth (citing Coleridge) articulates his search "to renew my stock of metaphors" through transfusions from recent scientific theory in his (and my own) admittedly nonscientific understanding.
14. Lefebvre, *The Production of Space*, 339.
15. Ibid.
16. Charles Guignon, Introduction to *The Cambridge Companion to Heidegger*, ed. Charles Guignon (Cambridge: Cambridge Univ. Press, 1993), 10.
17. David Couzens Hoy, "Heidegger and the Hermeneutic Turn," *The Cambridge Companion to Heidegger*, 172, 189.
18. These few paragraphs on Heidegger and his *Being and Time* are intended to suggest routes of further inquiry, not to propose any conclusive statements about Heidegger's philosophical stance on knowledge and meaning. As Hoy himself notes, the metaphor of the "circle of understanding"—in which self and world interact—can be interpreted in different ways, as either "static" or as "a dynamic process." There is, Hoy suggests, still a "coherent and unified" structure to this process which can be discerned; as I am suggesting, field dynamics are describable, even if every individual field is open to fluctuating accounts. Ibid., 190–91.
19. Lefebvre, *The Production of Space*, 362.
20. Ibid., 353.
21. Foucault, "Of Other Spaces," 23.

Selected Bibliography

Albanese, Catherine. *Corresponding Motion: Transcendental Religion and the New America*. Philadelphia: Temple Univ. Press, 1977.

Apollonio, Umbro, ed. *Futurist Manifestos*. Translated by Robert Brain et al. London: Thames and Hudson, 1973.

Bachelard, Gaston. *The Poetics of Space*. Edited and Translated by Maria Jolas. Boston: Beacon, 1994.

Barth, John. *Chimera*. New York: Fawcett Crest, 1972.

———. *The Floating Opera*. Rev. ed. New York: Doubleday, 1967.

———. "4 1/2 Lectures: The Stuttgart Seminars on Postmodernism, Chaos Theory, and the Romantic Arabesque." *Further Fridays: Essays, Lectures, and Other Nonfiction 1984–94*. New York: Little, 1995.

———. *The Last Voyage of Somebody the Sailor*. Boston: Little, Brown, 1991.

———. "The Literature of Replenishment: Postmodernist Fiction." *The Atlantic* 245 (January 1980): 65–71.

———. *Lost in the Funhouse: Fiction for Print, Tape, Live Voice*. Rev. ed. New York: Anchor, 1988.

———. *Once Upon a Time: a Floating Opera*. New York: Little, Brown, 1994.

———. *On With the Story*. New York: Little, Brown, 1996.

Bell, Steven. "Literature, Self-Consciousness, and Writing: The Example of Barth's *Lost in the Funhouse*." *The International Fiction Review* 11, no. 2 (summer 1984): 90–95.

Bloomer, Kent and Charles Moore. *Body, Memory, and Architecture*. New Haven: Yale Univ. Press, 1977.

Boccioni, Umberto, et al. "Futurist Painting: Technical Manifesto 1910." *Futurist Manifestos*. Edited by Umbro Apollonio. Translated by Robert Brain et al., pp. 27–31. London: Thames and Hudson, 1973.

———. "The Exhibitors to the Public 1912." *Futurist Manifestos*. Edited by Umbro Apollonio. Translated by Robert Brain et al., 45–50. London: Thames and Hudson, 1973.

———. "Plastic Dynamism 1913." *Futurist Manifestos*. Edited by Umbro Apollonio. Translated by Robert Brain et al., 92–95. London: Thames and Hudson, 1973.

Boone, Joseph Allen. "Delving and Diving for Truth: Breaking Through to Bottom in Thoreau's *Walden*." *Critical Essays on Henry David Thoreau's Walden*. Edited by Joel Myerson. Boston: G. K. Hall, 1988.

Borges, Jorge Luis. *Labyrinths*. Enl. ed. New York: New Directions, 1964.

Bowen, Zack. *A Reader's Guide to John Barth*. New York: Greenwood, 1994.

Burbick, Joan. *Thoreau's Alternative History: Changing Perspectives on Nature, Culture, and Language*. Philadelphia: Univ. of Pennsylvania Press, 1987.

Carmichael, Thomas. "Buffalo / Baltimore, Athens / Dallas: John Barth, Don DeLillo and the Cities of Postmodernism." *Canadian Review of American Studies* 22, no. 2 (fall, 1991): 241–49.

Chandler, Marilyn. *Dwelling in the Text: Houses in American Fiction*. Berkeley and Los Angeles: Univ. of California Press, 1991.

Cohen, Milton. *PoetandPainter: The Aesthetics of E. E. Cummings's Early Work*. Detroit: Wayne State Univ. Press, 1987.

Cooper, Clare. "The House as Symbol of the Self." *Designing for Human Behavior: Architecture and the Behavioral Sciences*. Edited by Jon Lang, 130–46. Stroudsburg, Pa.: Dowden, Hutchinson, and Ross, 1974.

Corddry, Mary. *City on the Sand: Ocean City, Maryland, and the People Who Built It*. Centreville, Maryland: Tidewater, 1991.

Cummings, E. E. *Complete Poems, 1904–1962*. Edited by George Firmage. Rev. ed. New York: Liveright, 1991.

———. *EIMI*. 2d ed. New York: Sloan, 1933.

———. *The Enormous Room*. Edited by George Firmage, 5–11. New York: Liveright, 1978.

———. *i six nonlectures*. Cambridge: Harvard Univ. Press, 1959.

———. Introduction to *The Enormous Room*. 3rd ed. New York: Random, 1949.

———. "The New Art." *E. E. Cummings: A Miscellany Revised*. Edited by George Firmage, New York: October House, 1965, 5–11.

Eddy, Arthur. *Cubists and Post-Impressionism*. Chicago: A. C. Clurg, 1914.

Einstein, Albert. Foreword to *Concepts of Space: the History of Theories of Space in Physics*, by Max Jammer. 3rd ed. New York: Dover, 1993.

Eliade, Mircea. *The Sacred and the Profane: The Nature of Religion*. Translated by William R. Trask. New York: Harvest, 1959.

Emerson, Ralph Waldo. *Essays: First and Second Series*. Edited by Joel Porte. New York: Vintage, 1990.

———. *Selections from Ralph Waldo Emerson*. Edited by Stephen E. Whicher. Boston: Houghton, 1957.

Foucault, Michel. *The Archaeology of Knowledge*. Translated by A. M. Sheridan Smith. New York: Pantheon, 1972.

———. "Of Other Spaces." Translated by Jay Miskowiec. *Diacritics: A Review of Contemporary Criticism* 6, no. 1 (spring 1986): 24–29.

———. *Power / Knowledge: Selected Interviews and Other Writings, 1972–1977*. Edited by Colin Gordon. Translated by Colin Gordon et. al. New York: Pantheon, 1980.

Friedman, Norman. *e.e. cummings: The Growth of a Writer*. Carbondale: Southern Illinois Univ. Press, 1964.

Frothingham, Octavius Brooks. *Transcendentalism in New England*. 1876. Reprint, New York: Harper, 1959.

Fryer, Judith. *Felicitous Space: The Imaginative Structures of Edith Wharton and Willa Cather*. Chapel Hill: Univ. of North Carolina Press, 1986.

Genette, Gérard. *Figures of Literary Discourse*. Translated by Alan Sheridan. New York: Columbia Univ. Press, 1982.

Guignon, Charles. Introduction to *The Cambridge Companion to Heidegger*. Edited by Charles Guignon, 1–41. Cambridge: Cambridge Univ. Press, 1993.

Harbison, Robert. *Eccentric Spaces: a Journey Through Real and Imaginary Worlds*. New York: Ecco Press, 1994.

Harding, Walter. *The Days of Henry Thoreau: A Biography*. Rev. and enl. ed. New York: Dover, 1982.

Harland, Richard. *Beyond Superstructuralism: The Syntagmatic Side of Language*. New York: Routledge, 1993.

Hawthorne, Nathaniel. "The Old Manse." *Mosses From an Old Manse*, 3–35. Ohio State Univ. Press, 1974.

Heidegger, Martin. "Building Dwelling Thinking." *Poetry, Language, Thought*. Translated by Albert Hofstadter. 145–61. New York: Harper, 1971.

Hershberger, Robert. "Predicting the Meaning of Architecture." *Designing for Human Behavior: Architecture and the Behavioral Sciences*. Edited by Jon Lang, Charles Burnette, Wlater Moleski, and David Vachon, 147–156. Stroudsburg, Pa.: Dowden, Hutchinson and Ross, 1974.

Hoy, David Couzens. "Heidegger and the Hermeneutic Turn." *The Cambridge Companion to Heidegger*. Edited by Charles Guignon, 170–94. Cambridge: Cambridge Univ. Press, 1993.

Hurley, George and Suzanne. *Ocean City: a Pictorial History*. Virginia Beach: Donning, 1979.

Huxtable, Ada Louise. "The New Architecture." *New York Review of Books* (April 6, 1995): 18–21.

Jakobsen, Roman. "Closing Statement: Linguistics and Poetics." *Style in Language*. Edited by Thomas A. Sebeok, 350–77. New York: Technology Press of Massachusetts, 1960.

James, Henry. "The Jolly Corner." *The Novels and Tales of Henry James*. Vol. 17. New York: Scribner, 1937.

Jameson, Fredric. *Postmodernism, or, The Cultural Logic of Late Capitalism*. Durham: Duke Univ. Press, 1995.

———. *The Prison-House of Language: A Critical Account of Structuralism and Russian Formalism*. Princeton: Princeton Univ. Press, 1972.

Jammer, Max. *Concepts of Space: The History of Theories of Space in Physics*. 3rd ed. New York: Dover, 1993.

Jarmusch, Jim. *Dead Man*. 1995. Black and white film, 120 minutes.

Kapleau, Roshi Philip. *The Three Pillars of Zen: Teaching, Practice, and Enlightenment*. Rev. ed. New York: Doubleday, 1989.

Kaku, Michio. *Hyperspace: A Scientific Odyssey Through Parallel Universes, Time Warps, and the 10th Dimension*. New York: Oxford Univ. Press, 1994.

Kasson, John. *Amusing the Million: Coney Island at the Turn of the Century*. New York: Hill and Wang, 1978.

Kennedy, Richard. *E. E. Cummings revisited*. New York: Twayne, 1994.

Keogh, Peter. "Home and Away." *Sight and Sound* 2, no. 4 (August 1992): 8–9.

Kern, Stephen. *The Culture of Time and Space: 1880–1918*. Cambridge: Harvard Univ. Press, 1983.

King, Ross. *Emancipating Space: Geography, Architecture, and Urban Design*. New York: Guilford, 1996.

Lang, Jon. "Theories of Perception and 'Formal' Design." In *Designing for Human Behavior: Architecture and the Behavioral Sciences*. Jon Lang, Charles Burnette, Wlater Moleski, and David Vachon, eds. 98–110. Stroudsburg, Pa.: Dowden, Hutchinson and Ross, 1974.

Lefebvre, Henri. *The Production of Space*. Translated by Donald Nicholson-Smith. Cambridge: Blackwell, 1991.

Lindsay, Alan. *Death in the FUNhouse: John Barth and Poststructuralist Aesthetics*. New York: Peter Lang, 1995.

Lista, Giovanni. *Futurism*. Translated by Charles Lynn Clark. New York: Universe Books, 1986.

Marinetti, Filippo Tommaso. "Technical Manifesto of Futurist Literature (1912)." *Marinetti: Selected Writings*. Edited by R. W. Flint. Translated by R. W. Flint and Arthur Coppotelli, 84–93. New York: Farrar, Straus and Giroux, 1971.

Marvell, Andrew. *Andrew Marvell*. Oxford: Oxford Univ. Press, 1990.

Massumi, Brian. "Sensing the Virtual, Building the Insensible." *Architectural Design: Hypersurface Architecture* 68, no. 5/6. (May–June 1998): 16–31.

McHale, Brian. *Postmodernist Fiction*. New York: Methuen, 1987.

Mott, Christopher. "The Cummings Line on Race." *Spring, A Journal of the E. E. Cummings Society*. New Series, no. 4 (October 1995): 71–75.

Nye, Russel. "Eight Ways of Looking at an Amusement Park." *Journal of Popular Culture* 15, no. 1 (summer 1981): 63–75.

Olson, Charles. *Call Me Ishmael: A Study of Melville*. New York: Grove, 1947.

Pearson, Michael and Colin Richards, "Ordering the World: Perceptions of Architecture, Space and Time." *Architecture and Order: Approaches to Social Space*, Michael Pearson and Colin Richards, eds. 1–37. New York: Routledge, 1994.

Perrella, Stephen. "Hypersurface Theory: Architecture><Culture." *Architectural Design: Hypersurface Architecture* 68. no. 5/6. (May–June, 1998): 7–15.

Perrella, Stephen and Rebecca Carpenter. "The Möbius House Study." *Architectural Design: Hypersurface Architecture* 68, no. 5/6. (May–June, 1998): 80–83.

Pinker, Steven. *The Language Instinct*. New York: William Morrow, 1994.

Poe, Edgar Allan. *The Complete Tales and Poems of Edgar Allan Poe*. New York: Vintage, 1975.

Poirier, Richard. *A World Elsewhere: The Place of Style in American Literature*. 2d ed. Univ. of Wisconsin Press, 1985.

Richardson, Robert Jr. *Henry Thoreau: A Life of the Mind*. Berkeley and Los Angeles: Univ. of California Press, 1986.

Soja, Edward. *Postmodern Geographies: The Reassertion of Space in Critical Social Theory*. New York: Verso, 1989.

Stea, David. "Architecture in the Head: Cognitive Mapping." *Designing for Human Behavior: Architecture and the Behavioral Sciences*. Jon Lang, Charles

Burnette, Wlater Moleski, and David Vachon, eds., 157–68. Stroudsburg, Pa.: Dowden, Hutchinson and Ross, 1974.

Stein, Gertrude. *Tender Buttons: Objects, Food, Rooms*. New York: Claire Marie, 1914.

———. *How to Write*. New York: Dover, 1975.

Stewart, Lynn. "Bodies, Visions, and Spatial Politics: A Review Essay on Henri Lefebvre's *The Production of Space*." *Environment and Planning D: Society and Space* 13, no. 5 (October 1995): 609–18.

Tales from the Thousand and One Nights. Translated by N. J. Dawood. New York: Penguin, 1973.

Tanner, Tony. *The City of Words: American Fiction, 1950–1970*. New York: Harper and Row, 1971.

Tharpe, Jack. *John Barth: The Comic Sublimity of Paradox*. Carbondale: Southern Illinois Univ. Press, 1974.

Thompson, Ben. Review of *Dead Man*. *Sight and Sound* 6, no. 7 (July 1996): 41–42.

Thoreau, Henry David. *Henry David Thoreau*. Edited by Robert F. Sayre. New York: Library of America, 1985.

———. *The Journal of Henry David Thoreau*. Edited by Bradford Torrey and Francis Allen. Salt Lake City: Peregrine, 1984.

Van de Ven, Cornelis. *Space in Architecture: The Evolution of a New Idea in the Theory and History of the Modern Movements*. 3rd ed. Assen, Netherlands: Van Gorcum, 1987.

Vitanza, Victor. "The Novelist as Topologist: John Barth's *Lost in the Funhouse*." *Texas Studies in Literature and Language* 19, no. 1 (Spring 1977): 83–97.

Walkiewicz, E. P. *John Barth*. New York: Twayne, 1986.

Zevi, Bruno. *Architecture as Space: How to Look at Architecture*. Translated by Milton Gendel. Edited by Joseph A. Barry. Rev. ed. New York: Horizon, 1974.

Index